ENERGY POLICIES, POLITICS AND PRICES

U.S. ENERGY AND THE ENVIRONMENT: AN OVERVIEW AND COMPARATIVE ANALYSIS

ENERGY POLICIES, POLITICS AND PRICES

Additional books in this series can be found on Nova's website at:

https://www.novapublishers.com/catalog/index.php?cPath=23_29&seriesp=Energy%20Policies%2C%20Politics%20and%20Prices&sort=2a&page=2

Additional E-books in this series can be found on Nova's website at:

https://www.novapublishers.com/catalog/index.php?cPath=23_29&seriespe=Energy+Policies%2C+Politics+and+Prices

ENERGY POLICIES, POLITICS AND PRICES

U.S. ENERGY AND THE ENVIRONMENT: AN OVERVIEW AND COMPARATIVE ANALYSIS

ROLAND H. TERRISON
EDITOR

Nova Science Publishers, Inc.
New York

For permission to use material from this book please contact us:
Telephone 631-231-7269; Fax 631-231-8175
Web Site: http://www.novapublishers.com

LIBRARY OF CONGRESS CATALOGING-IN-PUBLICATION DATA
U.S. energy and the environment : an overview and comparative analysis /
editor, Roland H. Terrison.
p. cm.
Includes bibliographical references and index.
ISBN 978-1-61668-017-6 (hardcover)

Published by Nova Science Publishers, Inc. ✢ New York

Contents

Preface		**vii**
Chapter 1	The Manhattan Project, the Apollo Program, and Federal Energy Technology R&D Programs: A Comparative Analysis *Deborah D. Stine*	**1**
Chapter 2	Nuclear Waste Disposal: Alternatives to Yucca Mountain *Mark Holt*	**15**
Chapter 3	U.S. Energy: Overview and Selected Facts and Numbers *Carl E. Behrens and Carol Glover*	**49**
Chapter 4	Biochar: Examination of an Emerging Concept to Mitigate Climate Change *Kelsi S. Bracmort*	**81**
Chapter 5	Alternative Fuels and Advanced Technology Vehicles: Issues in Congress *Brent D. Yacobucci*	**93**
Chapter 6	Carbon Tax and Greenhouse Gas Control: Options and Considerations for Congress *Jonathan L. Ramseur and Larry Parker*	**117**
Chapter 7	Environmental Quality Incentives Program (EQIP): Status and Issues *Megan Stubbs*	**179**

Chapter Sources 191

Index 193

PREFACE

Energy supplies and prices are major economic factors in the U.S., and energy markets are volatile and unpredictable. As an aid in policy making, this book presents a current and historical view of the supply and consumption of various forms of energy. With the downturn in the world economy and a consequent decline in consumption, prices collapsed, but the dependence continues as a potential problem. Other energy supplies discussed herein are: The Manhattan Project, the Apollo Program, and the Federal Technology R&D Programs; Nuclear Waste Disposal-Alternatives to Yucca Mountain; Biochar-An Examination of an Emerging Concept to Mitigate Climate Change; Alternative Fuels and Advanced Technology Vehicles; Carbon Tax and Greenhouse Gas Control; and the Environmental Quality Incentives Program (EQIP).

Chapter 1 - Some policymakers have concluded that the energy challenges facing the United States are so critical that a concentrated investment in energy research and development (R&D) should be undertaken. The Manhattan project, which produced the atomic bomb, and the Apollo program, which landed American men on the moon, have been cited as examples of the success such R&D investments can yield. Investment in federal energy technology R&D programs of the 1970s, in response to two energy crises, have generally been viewed as less successful than the earlier two efforts. This chapter compares and contrasts the three initiatives.

In 2008 dollars, the cumulative cost of the Manhattan project over 5 fiscal years was approximately $22 billion; of the Apollo program over 14 fiscal years, approximately $98 billion; of post-oil shock energy R&D efforts over 35 fiscal years, $118 billion. A measure of the nation's commitments to the programs is their relative shares of the federal outlays during the years of peak funding: for the Manhattan program, the peak year funding was 1% of federal outlays; for the Apollo program, 2.2%; and for energy technology R&D programs, 0.5%. Another measure of the commitment is their relative shares of the nation's gross domestic product (GDP) during the peak years of funding: for the Manhattan project and the Apollo program, the peak year funding reached 0.4% of GDP, and for the energy technology R&D programs, 0.1%.

Besides funding, several criteria might be used to compare these three initiatives including perception of the program or threat, goal clarity, and the customer of the technology being developed. By these criteria, while the Manhattan project and the Apollo program may provide some useful analogies for thinking about an energy technology R&D initiative, there are fundamental differences between the forces that drove these historical R&D success stories and the forces driving energy technology R&D today. Critical differences include (1) the ability to transform the program or threat into a concrete goal, and (2) the use to which the technology would be put. On the issue of goal setting, for the Manhattan project, the response to the threat of enemy development of a nuclear bomb was the goal to construct a bomb; for the Apollo program, the threat of Soviet space dominance was translated into a specific goal of landing on the moon. For energy, the response to the problems of insecure oil sources and high prices has resulted in multiple, sometimes conflicting, goals. Regarding use, both the Manhattan project and the Apollo program goals pointed to technologies primarily for governmental use with little concern about their environmental impact; for energy, in contrast, the hoped-for outcome depends on commercial viability and mitigation of environmental impacts from energy use.

Although the Manhattan project and the Apollo program may provide some useful analogies for funding, these differences may limit their utility regarding energy policy. Rather, energy technology R&D has been driven by at least three not always commensurate goals—resource and technological diversity, commercial viability, and environmental protection—which were not goals of the historical programs.

Some policymakers have concluded that the energy challenges facing the United States are so critical that a concentrated investment in energy research and development (R&D) should be undertaken. The Manhattan project, which produced the atomic bomb, and the Apollo program, which landed American men on the moon, have been cited as examples of the success such R&D investments can yield. Investment in federal energy technology R&D programs of the 1970s, in response to two energy crises, have generally been viewed as less successful than the earlier two efforts. This chapter compares and contrasts the goals of, and the investments in, the three initiatives, which may provide useful insights for Congress as it assesses and debates the nation's energy policy.

Chapter 2 - designated Yucca Mountain, NV, as the nation's sole candidate site for a permanent high-level nuclear waste repository in 1987, following years of controversy over the site-selection process. Over the strenuous objections of the State of Nevada, the Department of Energy (DOE) submitted a license application for the proposed Yucca Mountain repository in June 2008 to the Nuclear Regulatory Commission (NRC). During the 2008 election campaign, now-President Obama lent support to Nevada's fight against the repository, contending in an issue statement that he and now-Vice President Biden "do not believe that Yucca Mountain is a suitable site."

Under the current nuclear waste program, DOE hopes to begin transporting spent nuclear fuel and other highly radioactive waste to Yucca Mountain by 2020. That schedule is 22 years beyond the 1998 deadline established by the Nuclear Waste Policy Act (NWPA). Because U.S. nuclear power plants will continue to generate nuclear waste after a repository opens, DOE estimates that all waste could not be removed from existing reactors until about 2066 even under the current Yucca Mountain schedule. Not all the projected waste could be disposed of at Yucca Mountain, however, unless NWPA's current limit on the repository's capacity is increased.

If the Obama Administration decides to halt the Yucca Mountain project, it has a variety of tools available to implement that policy. Although the President cannot directly affect NRC proceedings, the Secretary of Energy could withdraw the Yucca Mountain license application under NRC rules. The President could also urge Congress to cut or eliminate funding for the Yucca Mountain project, and propose legislation to restructure the nuclear waste program.

Abandonment of Yucca Mountain would probably further delay the federal government's removal of nuclear waste from reactor sites and therefore increase the government's liabilities for missing the NWPA deadline. DOE estimates that such liabilities will reach $11 billion even if Yucca Mountain opens as currently planned. DOE's agreements with states to remove defense-related high-level waste could also be affected. If the Yucca Mountain project were halted without a clear alternative path for waste management, the licensing of proposed new nuclear power plants could be affected as well. NRC has determined that waste can be safely stored at reactor sites for at least 30 years after a reactor shuts down and is proposing to extend that period to 60 years. While that proposal would allow at least 100 years for waste to remain at reactor sites (including a 40-year reactor operating period), NRC's policy is that new reactors should not be licensed without "reasonable confidence that the wastes can and will in due course be disposed of safely."

Current law provides no alternative repository site to Yucca Mountain, and it does not authorize DOE to open temporary storage facilities without a permanent repository in operation. Without congressional action, therefore, the default alternative to Yucca Mountain would be indefinite on- site storage of nuclear waste at reactor sites and other nuclear facilities. Private central storage facilities can also be licensed under current law; such a facility has been licensed in Utah but its operation has been blocked by the Department of the Interior.

Congress has considered legislation repeatedly since the mid-1990s to authorize a federal interim storage facility for nuclear waste but none has been enacted. Reprocessing of spent fuel could reduce waste volumes and long-term toxicity, but such facilities are costly and raise concerns about the separation of plutonium that could be used in nuclear weapons. Storage and reprocessing would still eventually require a permanent repository, and a search for a new repository site would need to avoid the obstacles that have hampered previous U.S. efforts.

Chapter 3 - Energy supplies and prices are major economic factors in the United States, and energy markets are volatile and unpredictable. Thus, energy policy has been a recurring issue for Congress since the first major crisis in the 1970s. As an aid in policy making, this chapter presents a current and historical view of the supply and consumption of various forms of energy.

The historical trends show petroleum as the major source of energy, rising from about 38% in 1950 to 45% in 1975, then declining to about 40% in response to the energy crisis of the 1970s. Significantly, the transportation sector has been and continues to be almost completely dependent on petroleum, mostly gasoline. The importance of this dependence on the volatile world oil market was revealed over the past five years as perceptions of impending inability of the industry to meet increasing world demand led to relentless increases in the prices of oil and gasoline. With the downturn in the world economy and a consequent decline in consumption, prices collapsed, but the dependence on imported oil continues as a potential problem.

Natural gas followed a similar pattern at a lower level, increasing its share of total energy from about 17% in 1950 to more than 30% in 1970, then declining to about 20%. Consumption of coal in 1950 was 35% of the total, almost equal to oil, but it declined to about 20% a decade later and has remained at about that proportion since then. Coal currently is used almost exclusively for electric power generation.

Nuclear power started coming online in significant amounts in the late 1960s. By 1975, in the midst of the oil crisis, it was supplying 9% of total electricity generation. However, increases in capital costs, construction delays, and public opposition to nuclear power following the Three Mile Island accident in 1979 curtailed expansion of the technology, and many construction projects were cancelled. Continuation of some construction increased the nuclear share of generation to 20% in 1990, where it remains currently. The first new reactor license applications in nearly 30 years were recently submitted, but no new plants are currently under construction or on order.

Construction of major hydroelectric projects has also essentially ceased, and hydropower's share of electricity generation has gradually declined, from 30% in 1950 to 15% in 1975 and less than 10% in 2000. However, hydropower remains highly important on a regional basis.

Renewable energy sources (except hydropower) continue to offer more potential than actual energy production, although fuel ethanol has become a significant factor in transportation fuel, and wind power has recently grown rapidly. Conservation and energy efficiency have shown significant gains over the past three decades and offer encouraging potential to relieve some of the dependence on imports that has caused economic difficulties in the past, as well as the present.

After an introductory overview of aggregate energy consumption, this chapter presents detailed analysis of trends and statistics regarding specific energy sources: oil, electricity, natural gas, coal and renewable energy. A section on trends in energy efficiency is also presented.

Chapter 4 - Biochar is a charcoal produced under high temperatures using crop residues, animal manure, or any type of organic waste material. Biochar looks very similar to potting soil. The combined production and use of biochar is considered a carbon-negative process, meaning that carbon is removed from the atmosphere and will not be released into the atmosphere at a later time.

Biochar has multiple potential environmental benefits, foremost the potential to sequester carbon in the soil for hundreds to thousands of years at an estimate. Studies suggest that crop yields can increase as a result of applying biochar as a fertilizer to the soil. Some contend that biochar has value as an immediate climate change mitigation strategy. Scientific experiments suggest that greenhouse gas emissions are reduced significantly with biochar application to crop fields.

Obstacles that may stall rapid adoption of biochar production systems include technology costs, system operation and maintenance, feedstock availability, and biochar handling. Biochar research and development is in its infancy. Nevertheless, interest in biochar as a multifaceted solution to agricultural and natural resource issues is growing at a rapid pace both nationally and internationally.

Past Congresses have proposed numerous climate change bills, many of which do not directly address mitigation and adaptation technologies at developmental stages like biochar. However, biochar may equip agricultural and forestry producers with numerous revenue-generating products: carbon offsets, fertilizer, and energy. A clearly defined policy medium that supports this technology has yet to emerge (e.g., soil conservation, alternative energy, climate change).

This chapter briefly describes biochar, its potential advantages and disadvantages, legislative support, and research and development activities underway in the United States and abroad.

Chapter 5 - Alternative fuels and advanced technology vehicles are seen by proponents as integral to improving urban air quality, decreasing dependence on foreign oil, and reducing emissions of greenhouse gases. However, major barriers—especially economics—currently prevent the widespread use of these fuels and technologies. Because of these barriers, and the potential benefits, there is continued congressional interest in providing incentives and other support for their development and commercialization.

Alternative fuels and advanced technology vehicles have been addressed early in the 111th Congress, as both the House and Senate versions of the American Recovery and Reinvestment Act of 2009 (H.R. 1) contained provisions supporting their development and deployment. While some of these provisions were removed in conference, the final version still contains provisions for tax incentives, federal grants and loans, and other federal support for alternative fuels and advanced vehicles.

The 111th Congress is likely to further discuss alternative fuels and advanced technology vehicles as it addresses other key topics. These include their role in any federal policy to address climate change, and their role in federal energy policy. The 111th Congress may also play an oversight role in the development of major regulations: the Environmental Protection Agency's implementation of the renewable fuel standard enacted in 2005, and expanded in 2007; the Department of Transportation's implementation of new fuel economy standards enacted in 2007; and the Department of Agriculture's implementation of a new Farm Bill enacted in 2008.

In the 110th Congress, alternative fuels and advanced technology vehicles received a good deal of attention, especially in discussions over U.S. energy security. In his January 24, 2007, State of the Union Address, President Bush called for the increased use of renewable and alternative motor fuels to 35 billion gallons annually by 2017. U.S. consumption was roughly five billion gallons in 2006. Therefore, such an initiative would mean a seven-fold increase in the use of these fuels over 11 years. On December 19, 2007, President Bush signed the Energy Independence and Security Act of 2007 (EISA, P.L. 110-140). EISA requires an increase in renewable fuel consumption to 9.0 billion gallons in 2008 and 36 billion gallons in 2022. Further within the 36- billion-gallon requirement, by 2022 the law mandates the use of 21 billion gallons of "advanced biofuels," defined as fuel derived from renewable biomass other than corn starch, with 50% lower lifecycle greenhouse gas emissions compared to petroleum fuels. The 110th Congress also enacted the Food, Conservation, and Energy Act of 2008 (2008 Farm Bill, P.L. 110-246)—which expanded and extended incentives for biofuels—as well as the Emergency Economic Stabilization Act of 2008 (EESA, P.L. 110-343)—which modified existing fuel tax credits, and established a tax credit for the purchase of plug-in vehicles.

Chapter 6 - Market-based mechanisms that limit greenhouse gas (GHG) emissions can be divided into two types: quantity control (e.g., cap-and-trade) and price control (e.g., carbon tax or fee). To some extent, a carbon tax and a cap-and-trade program would produce similar effects: Both are estimated to increase the price of fossil fuels, which would ultimately be borne by consumers, particularly households. Although there are multiple tools available to policymakers that could control GHG emissions—including existing statutory authorities—this chapter focuses on a carbon tax approach and how it compares to its more frequently discussed counterpart: cap-and-trade.

If policymakers had perfect information regarding the market, either a price (carbon tax) or quantity control (cap-and-trade system) instrument could be designed to achieve the same outcome. Because this market ideal does not exist, preference for a carbon tax or a cap-and-trade program ultimately depends on which variable one wants to control—emissions or costs. Although there are several design mechanisms that could blur the distinction, the gap between price control and quantity control can never be completely overcome.

A carbon tax has several potential advantages. With a fixed price ceiling on emissions (or their inputs—e.g., fossil fuels), a tax approach would not cause additional volatility in energy prices. A set price would provide industry with better information to guide investment decisions: e.g., efficiency improvements, equipment upgrades. Economists often highlight a relative economic efficiency advantage of a carbon tax, but this potential advantage rests on assumptions—about the expected costs and benefits of climate change mitigation—that are uncertain and controversial. Some contend that a carbon tax may provide implementation advantages: greater transparency, reduced administrative burden, and relative ease of modification.

The primary disadvantage of a carbon tax is that it would yield uncertain emission control. Some argue that the potential for irreversible climate change impacts necessitates the emissions certainty that is only available with a quantity-based instrument (e.g., cap-and-trade). Although it may present implementation challenges, policymakers could devise a tax program that allows some short-term emission fluctuations, while progressing toward a long-term emission reduction objective. Proponents argue that short-term emission fluctuations would be preferable to the price volatility that might be expected with a cap-and-trade system.

Although a carbon tax could possibly face more political obstacles than a cap-and-trade program, some of these obstacles may be based on misunderstandings of the differences between the two approaches or on assumptions that the tax would be set too low to be effective. Carbon tax proponents could possibly address these issues to some degree, but there remains considerable political momentum for a cap-and-trade program.

Chapter 7 - The Environmental Quality Incentives Program (EQIP) is a voluntary program that provides farmers with financial and technical assistance to plan and implement soil and water conservation practices. EQIP is the largest agriculture conservation financial assistance program for working lands. EQIP was first authorized in 1996 and was most recently revised by Section 2501 of the Food, Conservation, and Energy Act of 2008 (P.L. 110-246, the 2008 farm bill). It is a mandatory spending program (i.e., not subject to annual appropriations) and is administered by the U.S. Department of Agriculture's (USDA's) Natural Resources Conservation Service (NRCS). Funding is currently authorized to grow to $1.75 billion in FY20 12. Eligible land includes cropland, rangeland, pasture, non-industrial private forestland, and other land on which resource concerns related to agricultural production could be addressed through an EQIP contract.

With the 111th Congress facing tighter budget constraints, EQIP could face similar challenges with a potential reduction in mandatory funding levels and a continuing backlog of unfunded applications. A change in income limitations along with a new waiver created in the 2008 farm bill could also raise issues for the program. EQIP will also continue to face challenges in measuring environmental and program accomplishments.

In: U.S. Energy and the Environment... ISBN: 978-1-61668-017-6
Editors: Roland H. Terrison pp. 1-13

Chapter 1

THE MANHATTAN PROJECT, THE APOLLO PROGRAM, AND FEDERAL ENERGY TECHNOLOGY R&D PROGRAMS: A COMPARATIVE ANALYSIS

Deborah D. Stine

SUMMARY

Some policymakers have concluded that the energy challenges facing the United States are so critical that a concentrated investment in energy research and development (R&D) should be undertaken. The Manhattan project, which produced the atomic bomb, and the Apollo program, which landed American men on the moon, have been cited as examples of the success such R&D investments can yield. Investment in federal energy technology R&D programs of the 1970s, in response to two energy crises, have generally been viewed as less successful than the earlier two efforts. This chapter compares and contrasts the three initiatives.

In 2008 dollars, the cumulative cost of the Manhattan project over 5 fiscal years was approximately $22 billion; of the Apollo program over 14 fiscal years, approximately $98 billion; of post-oil shock energy R&D efforts over 35 fiscal years, $118 billion. A measure of the nation's commitments to the programs is their relative shares of the federal outlays during the years of peak funding: for the Manhattan program, the peak year funding was 1% of federal outlays; for the Apollo program, 2.2%; and for energy technology R&D programs, 0.5%. Another measure of the commitment is their relative shares of the nation's gross domestic product (GDP) during the peak years of funding: for the Manhattan project and the Apollo program, the peak year funding reached 0.4% of GDP, and for the energy technology R&D programs, 0.1%.

Besides funding, several criteria might be used to compare these three initiatives including perception of the program or threat, goal clarity, and the customer of the technology being developed. By these criteria, while the Manhattan project and the Apollo program may provide some useful analogies for thinking about an energy technology R&D initiative, there are fundamental differences between the forces that drove these historical R&D success stories and the forces driving energy technology R&D today. Critical differences include (1) the ability to transform the program or threat into a concrete goal, and (2) the use to which the technology would be put. On the issue of goal setting, for the Manhattan project, the response to the threat of enemy development of a nuclear bomb was the goal to construct a bomb; for the Apollo program, the threat of Soviet space dominance was translated into a specific goal of landing on the moon. For energy, the response to the problems of insecure oil sources and high prices has resulted in multiple, sometimes conflicting, goals. Regarding use, both the Manhattan project and the Apollo program goals pointed to technologies primarily for governmental use with little concern about their environmental impact; for energy, in contrast, the hoped-for outcome depends on commercial viability and mitigation of environmental impacts from energy use.

Although the Manhattan project and the Apollo program may provide some useful analogies for funding, these differences may limit their utility regarding energy policy. Rather, energy technology R&D has been driven by at least three not always commensurate goals—resource and technological diversity, commercial viability, and environmental protection—which were not goals of the historical programs.

Some policymakers have concluded that the energy challenges facing the United States are so critical that a concentrated investment in energy research and development (R&D) should be undertaken. The Manhattan project, which produced the atomic bomb, and the Apollo program, which landed American men on the moon, have been cited as examples of the success such R&D investments can yield. Investment in federal energy technology R&D programs of the 1970s, in response to two energy crises, have generally been viewed as less successful than the earlier two efforts. This chapter compares and contrasts the goals of, and the investments in, the three initiatives, which may provide useful insights for Congress as it assesses and debates the nation's energy policy.

THE MANHATTAN PROJECT

The Manhattan project took place from 1942 to 1946.[1] Beginning in 1939, some key scientists expressed concern that Germany might be building an atomic weapon and proposed that the United States accelerate atomic research in response. Following the Pearl Harbor attack in December 1941, the United States entered World War II. In January 1942, President Franklin D. Roosevelt gave secret, tentative approval for the development of an atomic bomb. The Army Corps of Engineers was assigned the task and set up the Manhattan Engineer District to manage the project. A bomb research and design laboratory was built at Los Alamos, New Mexico. Due to uncertainties regarding production effectiveness, two possible fuels for the reactors were produced with uranium enrichment facilities at Oak Ridge, Tennessee, and plutonium production facilities at Hanford, Washington. In December 1942, Roosevelt gave final approval to construct a nuclear bomb. A bomb using plutonium as fuel was successfully tested south of Los Alamos in July 1945. In August 1945, President Truman decided to use the bomb against Japan at two locations. Japan surrendered a few days after the second bomb attack. At that point, the Manhattan project was deemed to have fulfilled its mission, although some additional nuclear weapons were still assembled. In 1946, the civilian Atomic Energy Commission was established to manage the nation's future atomic activities, and the Manhattan project officially ended.

According to one estimate, the Manhattan project cost $2.2 billion from 1942 to 1946 ($22 billion in 2008 dollars), which is much greater than the original cost and time estimate of approximately $148 million for 1942 to 1944.[2] General Leslie Groves, who managed the Manhattan project, has written that Members of Congress who inquired about the project were discouraged by the Secretary of War from asking questions or visiting sites.[3] After the project was under way for over a year, in February 1944, War Department officials received essentially a "blank check" for the project from Congressional leadership who "remained completely in the dark" about the Manhattan project, according to Groves and other experts.[4]

THE APOLLO PROGRAM

The Apollo program, FY1960 to FY1973, encompassed 17 missions, including six lunar landings.[5] NASA was created in response to the Soviet launch of Sputnik in 1958 and began operation in 1959.[6] Although preliminary discussions regarding the Apollo program began in 1960, Congress did not decide to fund it until 1961 after the Soviet Union became the first nation to launch a human into space. The goals of the Apollo program were

- To land Americans on the Moon and return them safely to Earth;
- To establish the technology to meet other national interests in space;
- To achieve preeminence in space for the United States;
- To carry out a program of scientific exploration of the Moon; and
- To develop man's capability to work in the lunar environment.[7]

The program included a three-part spacecraft to take two astronauts to the Moon surface, support them while on the Moon, and return them to Earth.[8] Saturn rockets were used to launch this equipment. In July 1969, Apollo 11 achieved the goal of landing Americans on the Moon and returning them safely to Earth. The last lunar landing took place in December 1972.

The Apollo program was only one part of NASA's activities during this period. NASA's peak funding during the Apollo program occurred in FY1966 when its total funding was $4.5 billion (in current dollars), of which $3.0 billion went to the Apollo program.[9] According to NASA, the total cost of the Apollo program for FY1 960-FY 1973 was $19.4 billion ($97.9 billion in 2008 dollars).[10] The activities with the greatest cost were the Saturn V rockets ($6.4 billion in current dollars) followed by the Command and Service Modules ($3.7 billion), the Lunar Modules ($2.2 billion), and Manned Space Flight Operations ($1.6 billion).

ENERGY TECHNOLOGY RESEARCH AND DEVELOPMENT

The Arab oil embargo of 1973 (the "first" energy crisis) put energy policy on the national "agenda." At that time, Americans began to experience rapidly rising prices for fuel and related goods and services.[11] Until then, energy R&D had been focused on the development of nuclear power under the Atomic Energy Commission (AEC). After the Manhattan project ended, Congress had established the AEC to manage both civilian and military projects in the Atomic Energy Act of 1946 (P.L. 79-585).[12] In response to the energy crisis, Congress subsumed the AEC, including the Manhattan project facilities, and other energy programs, into the Energy Research and Development Administration (ERDA), which became the focus for federal energy technology R&D, and the Nuclear Regulatory Commission (NRC) as part of the Energy Reorganization Act of 1974 (P.L. 93-438).[13]

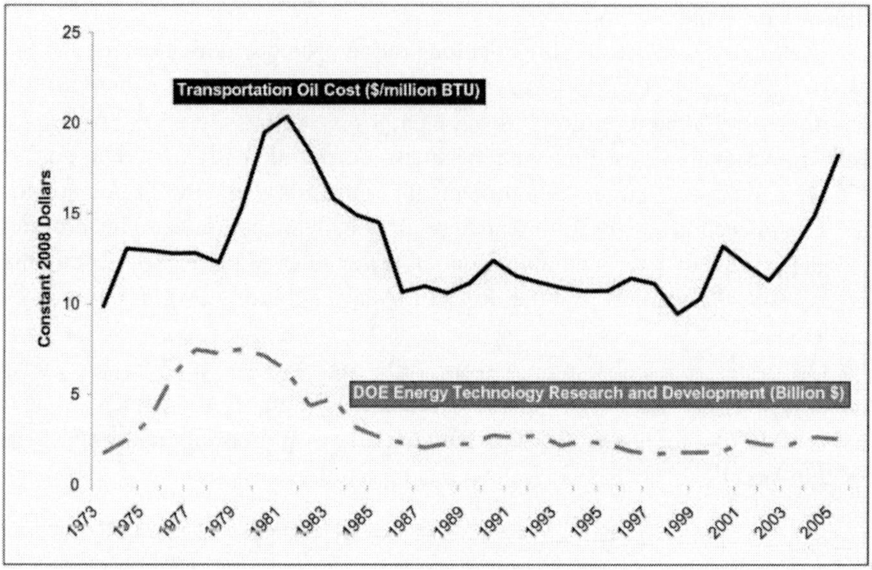

Source: Congressional Research Service. Transportation oil costs are from transportation/petroleum column of Table 3.4 Consumer Price Estimates for Energy by End-Use Sector, 1970-2005 in Energy Information Administration, Annual Energy Review 2007, Report No. DOE/EIA-0384(2007), Posted: June 23, 2008, at http://www.eia.doe.gov/aer/finan.html. DOE data is from CRS Report RS22858, Renewable Energy R&D Funding History: A Comparison with Funding for Nuclear Energy, Fossil Energy, and Energy Efficiency R&D, by Fred Sissine. Dollars for both transportation oil costs and DOE energy R&D were adjusted to 2008 dollars using the information from Bureau of Economic Analysis, Table 1.1.4. Price Indexes for Gross Domestic Product at *http://www.bea.gov/* bea/dn/ nipaweb/index.asp.

Figure 1. Comparison of Consumer Transportation Oil Cost and DOE Energy Technology R&D Funding, 1973-2005

In the Department of Energy Organization Act of 1977 (P.L. 95-91), Congress decided to combine the activities of ERDA with approximately 50 other energy offices and programs in a new Department of Energy (DOE), which began operations on October 1, 1977.[14] In 1979, the Iranian Revolution precipitated the "second" energy crisis that took place from 1978-1981. High oil prices and inflation lasted for several years. An ensuing recession curbed demand and oil prices fell markedly by 1986. The scale of funding for most of DOE's energy R&D programs dropped steadily during the 1980s (see **Figure 1 on Page 5).**

The large energy technology demonstration projects funded during the late 1970s and early 1980s were viewed by some as too elaborate and insufficiently linked to either existing energy research or the marketplace.[15] A well known example is the Synthetic Fuels Corporation (SFC).[16] The goal of SFC was to support large-scale projects that industry was unwilling to support due to the technical, environmental, or financial uncertainties. The program ended in 1986 due to a combination of lower energy prices, environmental issues, lack of support by the Reagan Administration, and administrative challenges.[17]

Oil prices rose substantially from 2004 to 2008, but funding for energy technology R&D has not so far increased as it did during the energy crisis of the late 1970s to early 1 980s. However, the appropriations process for FY2009 and FY20 10 has not yet been completed.

COMPARATIVE ANALYSIS OF THE MANHATTAN PROJECT, THE APOLLO PROGRAM, AND FEDERAL ENERGY TECHNOLOGY R&D

A general understanding of driving forces of and funding histories for the Manhattan project and Apollo program, and a comparison of these two initiatives to Department of Energy (DOE) energy technology R&D programs, may provide useful insights for Congress as it assesses and determines the nation's energy R&D policy. Four criteria that might be used to compare these programs are funding, perception of threat, goal clarity, and technology customer. Each is discussed in more depth below.

Funding

Table 1 provides a comparison of the total and annual average program costs for the Manhattan project, Apollo program, and federal energy technology R&D program since the first energy crisis. Annual average long-term (1974-2008) DOE energy technology R&D funding was approximately $3 billion (in 2008 constant dollars) as is the FY2008 budget and the FY2009 budget request.[18] In comparison, the annual average funding (in 2008 constant dollars) for the Manhattan project was $4 billion and for the Apollo program and the DOE energy technology program at its peak (1975-1980) was $7 billion.

Table 1. Cumulative and Annual Average Program Year Funding for the Manhattan

	Cumulative Funding (in billions of 2008 dollars)	Number of Fiscal Years	Annual Average Funding Per Program Year (in billions of 2008 dollars)	Percent of Federal Outlays During Year of Peak Funding	Percent of GDP during Year of Peak Funding
The Manhattan Project (1942- 1946)	$22	5	$4	1.0	0.4
The Apollo Program (1960-1973)	$98	14	$7	2.2	0.4
DOE Energy Technology Programs (1975- 1980)	$40	6	$7	0.5	0.1
[Peak Funding]					
DOE Energy Technology Program (1974- 2008)	$118	35	$3	0.5	0.1
[Long-Term					

Source: Congressional Research Service. Manhattan Project data: Richard G. Hewlett and Oscar E. Anderson, Jr., *A History of the United States Atomic Energy Commission: The New World, 1939/1946, Volume I*. Apollo program data: Richard Orloff, *Apollo By The Numbers: A Statistical Reference*, NASA SP-2000-4029, 2004 web update. DOE data: CRS Report RS22858, *Renewable Energy R&D Funding History: A Comparison with Funding for Nuclear Energy, Fossil Energy, and Energy Efficiency R&D*, by Fred Sissine. Federal Outlay and Gross Domestic Product (GDP) data: Office of Management and Budget, *Historical Tables, Budget of the United States Government FY2009*. Peak year of funding (in current dollars) for Manhattan project was 1946, for Apollo program, 1966, and for DOE Energy Technology R&D programs, 1980. The greatest annual funding (in constant dollars) for DOE energy technology programs took place from 1975-1980.

At the time of peak funding, the percentage of federal spending devoted to DOE energy technology R&D was half that of the Manhattan project, and one-fifth that of the Apollo program. From an overall economy standpoint, the percentage of the gross domestic product (GDP) spent on DOE energy technology R&D in the peak funding year was one-fourth that spent on either the Manhattan project or the Apollo program.

As shown in **Figure 2**, although cumulative funding for the DOE energy technology R&D program is greater than for the Manhattan project or the Apollo program, the annual funding for each of the historical programs was higher than that for energy technology R&D which occurred over a greater number of years. This is an important distinction: the Manhattan project and the Apollo program were specific and distinct funding efforts whereas the national energy R&D effort has been ongoing over a longer period of time. In all three cases, a rapid increase in funding was followed by a rapid decline.

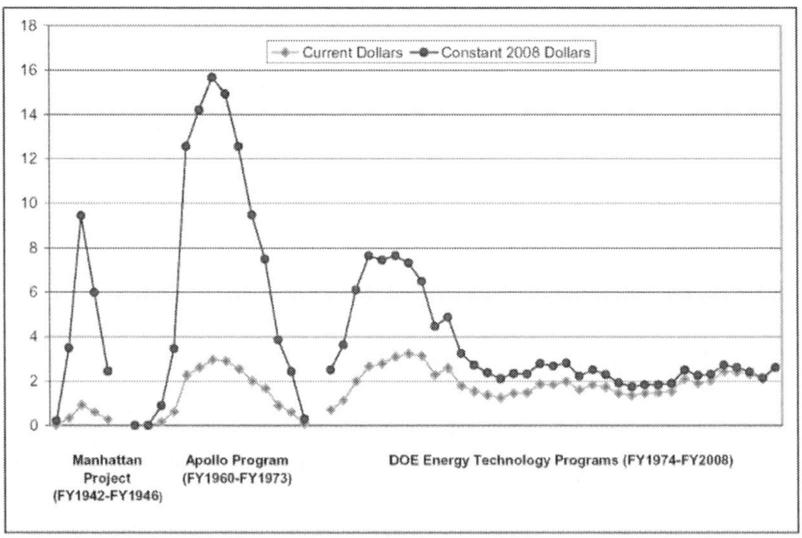

Source: Congressional Research Service. Manhattan Project data: Richard G. Hewlett and Oscar E. Anderson, Jr., *A History of the United States Atomic Energy Commission: The New World, 1939/1946, Volume I.* Apollo program data: Richard Orloff, *Apollo By The Numbers: A Statistical Reference*, NASA SP-2000-4029, 2004 web update. DOE data: CRS Report RS22858, *Renewable Energy R&D Funding History: A Comparison with Funding for Nuclear Energy, Fossil Energy, and Energy Efficiency R&D*, by Fred Sissine.

Figure 2. Annual Funding for Manhattan Project, Apollo Program, and DOE Energy Technology R&D Program

Threat Perception

The Manhattan project and Apollo project were both responses to perceived threats, which compelled policymaker support for these initiatives. The Manhattan project took place during World War II. Although the public might have been unaware of the potential threat of Germany's use of nuclear weapons and the Manhattan project, the President and Members of Congress could feel confident about public support for the war effort of which the Manhattan project was a part. Similarly, the Apollo program took place during the Cold War with the Union of Soviet Socialist Republics (USSR). When the USSR launched the Sputnik satellite and first man into space, the U.S. public felt threatened by the potential that the USSR might take leadership in the development of space flight technology, and potentially greater control of outer space. President Jimmy Carter said that

> With the exception of preventing war, this [energy crisis] is the greatest challenge that our country will face during our lifetime ... our decision about energy will test the character of the American people and the ability of the President and the Congress to govern this Nation. This difficult effort will be the 'moral equivalent of war,' except that we will be uniting our efforts to build and not to destroy."[19]

The threat to which investment in energy technology R&D responds, however, is largely economic rather than military. In addition, the threat posed by climate change, which is related to energy consumption, will likely be gradual and long-term.[20]

Goal Clarity

Another issue is the degree to which there is clarity and consensus on the program goal. The Manhattan project had a clear and singular goal—the creation of a nuclear bomb. For the Apollo program, the goal was also clear and singular—land American astronauts on the moon and return them safely to Earth. In the case of energy technology R&D, however, the overall goal of clean, affordable, and reliable energy is multi-faceted. While "energy independence" has from time to time been a rallying cry, energy technology R&D has in fact, been driven by at least three not always commensurate goals: resource and technological diversity, commercial viability, and environmental protection. To help reduce the risk of dependence on a single energy source, diversity of resources and energy technologies have always been seen as a goal of the energy R&D program. Second, unlike the Manhattan project or the Apollo program, the DOE energy technology R&D program seeks ultimately to be commercially viable. Third, the energy R&D program must meet environmental goals, including reducing the impact of energy-related activities on land, water, air, and climate change.

Technology Customer

Another comparison criterion is the customer for technologies that may result from the R&D. The government was the customer for both the Manhattan project and Apollo program. The private sector is the ultimate customer for any energy technology developed as a result of federal energy R&D programs. Therefore, the marketability of any technologies developed will be a key determinant of the degree to which the program is successful. Moreover, the inherent involvement of the private sector raises a number of issues related to the appropriateness of different government roles. Some believe that focusing R&D on one particular technology versus another may result in government, instead of the marketplace, picking "winners and losers."[21] Some experts believe that the most important driver for private sector deployment or commercialization is not the need for new technologies, but regulation or economic incentives.[22] Others, however, believe that without government support and intervention, the private sector is unlikely to conduct the R&D necessary to achieve the public goal of clean, affordable, and reliable energy, and that current technologies are insufficient to achieve this goal.[23]

Implications for Congress

When the Manhattan project and the Apollo program are used as analogies for future DOE energy technology R&D, the following points may be important to consider:

- To be equivalent in annual average funding, DOE energy technology R&D funding would need to increase from approximately $3 billion in FY2008 to at least $4 billion per program year to match the

Manhattan project funding, or $7 billion per program year to match Apollo program funding levels or DOE energy technology R&D funding at its peak. To be equivalent of peak year funding would require even greater increases. In terms of federal outlays, energy technology R&D funding would need to increase from 0.5% to 1% (Manhattan project) or 2.2 % (Apollo program) of federal outlays. As a percentage of GDP, this funding would need to increase from 0.1% to 0.4% of GDP (for both the Manhattan project and the Apollo program).

- Both the Manhattan project and the Apollo program had a singular and specific goal. For the Manhattan project, the response to the threat of enemy development of a nuclear bomb was the goal to construct a bomb; for the Apollo program, the threat of Soviet space dominance was translated into a specific goal of landing on the moon. For energy, however, the response to the problems of insecure oil sources and high prices has resulted in multiple, sometimes conflicting goals.
- Both the Manhattan project and the Apollo program goals pointed to technologies primarily for governmental use with little concern about their environmental impact; for energy, in contrast, the hoped for outcome depends on commercial viability and mitigation of the environmental impacts of the energy technologies developed.

Although the Manhattan project and the Apollo program may provide some useful analogies for funding, these differences may limit their utility regarding energy policy. Rather, energy technology R&D has been driven by at least three not always commensurate goals—resource and technological diversity, commercial viability, and environmental protection—which were not goals of the historical programs.

End Notes

[1] U.S. Department of Energy, Office of History and Heritage Resources, "The Manhattan Project: An Interactive History," http://www.cfo.doe.gov/me70/manhattan/1939-1942.htm. F.G. Gosling, *The Manhattan Project: Making the Atomic Bomb*, January 1999 edition (Oak Ridge, TN: Department of Energy).

[2] Richard G. Hewlett and Oscar E. Anderson, Jr., A *History of the United States Atomic Energy Commission: The New World, 1939/1946, Volume I,* (University Park, PA: The Pennsylvania State University Press, 1962). Appendix 2 provides the annual Manhattan project expenditures. These costs were adjusted to 2008 dollars using the price index for gross domestic product (GDP), available from the Bureau of Economic Affairs, National

Income and Product Accounts Table webpage, Table 1.1.4, at http://www.bea.gov/bea/dn/nipaweb/.

[3] Leslie R. Groves, *Now it Can be Told: The Story of the Manhattan Project* (New York: Harper & Brothers, 1962).

[4] Ibid., Kevin O'Neil, "Building the Bomb," Chapter 1 in Stephen I. Schwartz (ed.), *Atomic Audit: The Costs and Consequences of U.S. Nuclear Weapons Since 1940* (Washington, DC: Brookings Institution Press, 1998), pp. 58-59.

[5] Richard Orloff, National Aeronautics and Space Administration (NASA), *Apollo By The Numbers: A Statistical Reference*, NASA SP-2000-4029, 2004 web update at http://history.nasa.gov/SP-4029/Apollo_00_Welcome.htm. There is some difference of opinion regarding what activities comprised the Apollo program, and thus when it began and ended. For example, two different cost figures are provided on NASA's website. This is probably because some analysts include the first studies for Apollo, Skylab, and the use of Apollo spacecraft in the Apollo-Soyuz Test Project. The Orloff analysis includes the first studies of Apollo, but not Skylab (1973-74) or Soyuz (1975) activities. Another NASA analysis provides the cost as $25.4 billion, but provides no details as to how the cost were determined. See Roger D. Launius, NASA, *The Legacy of Project Apollo* at http://history.nasa.gov/ap11-35ann/legacy.html.

[6] For more information, see CRS Report RL34263, *U.S. Civilian Space Policy Priorities: Reflections 50 Years After Sputnik*, by Deborah D. Stine.

[7] NASA, Kennedy Space Center, "Project Apollo," webpage, at http://www-pao.ksc.nasa.gov/kscpao/history/apollo/ apollo.htm. A list of the top ten Apollo scientific discoveries as determined by the Smithsonian Institution is at http://www.nasm.si.edu/collections/imagery

[8] NASA, Kennedy Space Center, "Project Apollo," webpage, at http://www-pao.ksc.nasa.gov/kscpao/history/apollo/ apollo.htm. The three parts were the command module (CM), the crew's quarters and flight control section; the service module (SM) for the propulsion and spacecraft support systems (when together, the two modules are called CSM); and the lunar module (LM).

[9] The funding data is available at http://history.nasa.gov/SP. It is based on information in NASA, *The Apollo Spacecraft - A Chronology*, NASA Special Publication-4009, at http://www.hq.nasa.gov/office/pao/History/ SP-4009/contents.htm. This data is from Volume 4, Appendix 7, at http://www.hq.nasa.gov/office/pao/History/SP4009/v4app7.htm.

[10] Richard Orloff, *Apollo By The Numbers: A Statistical Reference*, NASA SP-2000-4029, 2004 web update, at http://history.nasa.gov/SP-4029/Apollo_00_Welcome.htm. The funding data is available at *http://history.nasa.gov/SP4029/Apollo_18-16_Apollo_Program_Budget_ Appropriations.htm*. It is based on information in NASA, *The Apollo Spacecraft - A Chronology*, NASA Special Publication-4009, at http://www.hq.nasa.gov/office/pao/History/SP-4009/ contents.htm.

[11] Energy Information Administration, "25th Anniversary of the 1973 Oil Embargo," at http://www.eia.doe.gov/emeu/ 25opec/anniversary.html.

[12] U.S. Department of Energy, Office of History and Heritage Resources, "The Manhattan Project: An Interactive History," "Civilian Control of Atomic Energy" webpage at http://www.cfo.doe.gov/me70/manhattan/ civilian_control.htm.

[13] Department of Energy, "Energy Research and Development Administration," webpage at http://www.ch.doe.gov/ html/site_info/energy_research.htm.

[14] Although DOE did not begin operating until 1977, the term "DOE Energy Technology R&D Program" in this chapter is defined as encompassing DOE programs funded beginning in 1977 as well as energy R&D activities that occurred prior to 1977 that were managed by the organizations it subsumed.

[15] See, for example, Bruce L.R. Smith, *American Science Policy Since World War II* (Washington, DC: Brookings Institution, 1990).

[16] The Synthetic Fuels corporation was established by the United States Synthetic Fuels Corporation Act of 1980 (P.L. 96-294), and its operation was discontinued by the Synthetic Fuels Corporation Act of 1985 (P.L. 99-272).

[17] National Academy of Sciences, *The Government Role in Civilian Technology: Building a New Alliance* (Washington, DC: National Academy Press 1992).

[18] For information on the DOE energy technology R&D budget, see CRS Report RL34448, *Federal Research and Development Funding: FY2009*, by John F. Sargent Jr. et al.

[19] President Jimmy Carter, "Address to the Nation on Energy," speech, April 18, 1977. A video and text is at http://millercenter.org/scripps/archive/speeches/detail/3398.

[20] CRS Report RL34513, *Climate Change: Current Issues and Policy Tools*, by Jane A. Leggett.

[21] For more discussion of this issue, see CRS Report RL33528, *Industrial Competitiveness and Technological Advancement: Debate Over Government Policy*, by Wendy H. Schacht.

[22] See for example, David Goldston, "Misspent Energy," *Nature*, 447:130, May 10, 2007 at http://www.nature.com/ nature/journal/v447/n7141/pdf/447130a.pdf.

[23] See, for example, The National Academies, Rising Above the Gathering Storm: Energizing and Employing America for a Brighter Economic Future (Washington, DC: National Academy Press, 2007).

In: U.S. Energy and the Environment… ISBN: 978-1-61668-017-6
Editors: Roland H. Terrison pp. 15-48

Chapter 2

NUCLEAR WASTE DISPOSAL: ALTERNATIVES TO YUCCA MOUNTAIN

Mark Holt

SUMMARY

Congress designated Yucca Mountain, NV, as the nation's sole candidate site for a permanent high-level nuclear waste repository in 1987, following years of controversy over the site-selection process. Over the strenuous objections of the State of Nevada, the Department of Energy (DOE) submitted a license application for the proposed Yucca Mountain repository in June 2008 to the Nuclear Regulatory Commission (NRC). During the 2008 election campaign, now-President Obama lent support to Nevada's fight against the repository, contending in an issue statement that he and now-Vice President Biden "do not believe that Yucca Mountain is a suitable site."

Under the current nuclear waste program, DOE hopes to begin transporting spent nuclear fuel and other highly radioactive waste to Yucca Mountain by 2020. That schedule is 22 years beyond the 1998 deadline established by the Nuclear Waste Policy Act (NWPA). Because U.S. nuclear power plants will continue to generate nuclear waste after a repository opens, DOE estimates that all waste could not be removed from existing reactors until about 2066 even under the current Yucca Mountain schedule. Not all the projected waste could be disposed of at Yucca Mountain, however, unless NWPA's current limit on the repository's capacity is increased.

If the Obama Administration decides to halt the Yucca Mountain project, it has a variety of tools available to implement that policy. Although the President cannot directly affect NRC proceedings, the Secretary of Energy could withdraw the Yucca Mountain license application under NRC rules. The President could also urge Congress to cut or eliminate funding for the Yucca Mountain project, and propose legislation to restructure the nuclear waste program.

Abandonment of Yucca Mountain would probably further delay the federal government's removal of nuclear waste from reactor sites and therefore increase the government's liabilities for missing the NWPA deadline. DOE estimates that such liabilities will reach $11 billion even if Yucca Mountain opens as currently planned. DOE's agreements with states to remove defense-related high-level waste could also be affected. If the Yucca Mountain project were halted without a clear alternative path for waste management, the licensing of proposed new nuclear power plants could be affected as well. NRC has determined that waste can be safely stored at reactor sites for at least 30 years after a reactor shuts down and is proposing to extend that period to 60 years. While that proposal would allow at least 100 years for waste to remain at reactor sites (including a 40-year reactor operating period), NRC's policy is that new reactors should not be licensed without "reasonable confidence that the wastes can and will in due course be disposed of safely."

Current law provides no alternative repository site to Yucca Mountain, and it does not authorize DOE to open temporary storage facilities without a permanent repository in operation. Without congressional action, therefore, the default alternative to Yucca Mountain would be indefinite on- site storage of nuclear waste at reactor sites and other nuclear facilities. Private central storage facilities can also be licensed under current law; such a facility has been licensed in Utah but its operation has been blocked by the Department of the Interior.

Congress has considered legislation repeatedly since the mid-1990s to authorize a federal interim storage facility for nuclear waste but none has been enacted. Reprocessing of spent fuel could reduce waste volumes and long-term toxicity, but such facilities are costly and raise concerns about the separation of plutonium that could be used in nuclear weapons. Storage and reprocessing would still eventually require a permanent repository, and a search for a new repository site would need to avoid the obstacles that have hampered previous U.S. efforts.

PROPOSALS FOR A NEW DIRECTION

Nevada's Yucca Mountain has been the sole candidate site for the nation's first permanent high- level nuclear waste repository since Congress singled it out in 1987 and halted consideration of any other location. After numerous delays, the Department of Energy (DOE), which was supposed to open a waste repository by 1998, submitted a repository license application for Yucca Mountain to the Nuclear Regulatory Commission (NRC) in June 2008. If NRC approves the license, DOE hopes to begin shipping nuclear waste to the repository by 2020.

The congressional decision to focus solely on Yucca Mountain was highly controversial and continues to face harsh criticism, particularly from the State of Nevada. Yucca Mountain opponents dispute DOE's determination that the site is suitable for long-term disposal of nuclear waste and call for fundamental change in the program. Nonetheless, the proposed Yucca Mountain repository has consistently maintained sufficient congressional support to continue moving forward. Congress explicitly rejected a Nevada "state veto" of the site in 2002, has blocked repeated efforts to halt the program's funding, and has not taken up any of numerous legislative proposals to delay the program or find a new site. President George W. Bush also steadfastly supported the Yucca Mountain project.

But Administration support for Yucca Mountain will apparently change under President Obama. In their campaign statement on nuclear energy policy, Obama and Vice President Biden laid out the following position:

> In terms of waste storage, Barack Obama and Joe Biden do not believe that Yucca Mountain is a suitable site. They will lead federal efforts to look for safe, long-term disposal solutions based on objective, scientific analysis. In the meantime, they will develop requirements to ensure that the waste stored at current reactor sites is contained using the most advanced dry- cask storage technology available.[1]

The Obama-Biden campaign statement on Yucca Mountain raises numerous questions about the future direction of U.S. nuclear waste policy. In particular, what type of long-term disposal solutions could be considered? Every option for handling nuclear waste ultimately requires a method of long-term isolation from the environment. If Yucca Mountain were rejected, it would appear that a new repository site search would need to be undertaken at some point. Given the criticism that DOE has drawn over its handling of the waste program, pressure may intensify for such a search to be handed over to a new organization entirely.

The current effort to develop a repository at Yucca Mountain began with the enactment of the Nuclear Waste Policy Act of 1982 (NWPA, P.L. 97-425), and opening a repository at a different location could take a long time as well, even if the process were started right away. During such an indefinite time period, how would the licensing of new nuclear power plants be affected? Would spent nuclear fuel and other highly radioactive waste remain at commercial reactors and other existing nuclear facilities, or would it be moved to centralized interim storage? Previous U.S. efforts to develop interim central nuclear waste storage facilities have drawn fierce opposition.

Since the 1970s, U.S. nuclear waste policy has been based on the "once through" fuel cycle, in which nuclear fuel is to be used once in a reactor and then permanently disposed of. The major alternative is the "closed" fuel cycle, in which spent nuclear fuel would be reprocessed into new fuel for advanced reactors or particle accelerators. Fast reactors or accelerators would destroy the longest-lived radioactive components of the fuel, leaving only relatively short-lived radioactive isotopes, which would decay to background levels within 1,000 years, for permanent disposal. Under that scenario, spent fuel could be stored at reprocessing facilities while awaiting its turn to be made into new fuel, and the relatively short life of the resulting waste could make it easier to site a permanent repository. However, the material for nuclear fuel that results from reprocessing (primarily plutonium) can also be used for nuclear explosives, raising concerns about nuclear weapons proliferation.

Because current law specifies that Yucca Mountain is the sole candidate site for a high-level waste repository, legislation would probably be needed if a major change in direction in the nuclear waste program is sought. But the Obama Administration does have authority under current law to withdraw the Yucca Mountain license application, propose reductions in the program's funding, and take other administrative actions to delay or halt the development of a repository at the Yucca Mountain site. This chapter discusses those options and the likely impact of indefinite delays in the waste program. It then discusses the mid- and long-term alternatives to the existing waste program, and finally reviews the history of U.S. efforts to site nuclear waste facilities.

BASELINE: CURRENT WASTE PROGRAM PROJECTIONS

DOE's latest schedules for nuclear waste shipments and projected costs under the existing program provide a baseline for analyzing Yucca Mountain alternatives. Although the planned opening of the Yucca Mountain repository is at least 22 years later than NWPA's 1998 deadline, and removing waste from existing storage sites would require many decades, a major redirection of the waste program would probably involve even longer time frames. Of course, there is no certainty that DOE will be able to meet its current schedules, or that the Yucca Mountain repository will receive a license from NRC under the current program. Moreover, policymakers could conclude that the benefits of redirecting the nuclear waste program now would outweigh the almost certain delays in developing a permanent repository and the increased costs of interim storage.

Under DOE's current schedule, about 400 metric tons of spent nuclear fuel would be shipped to Yucca Mountain from reactor sites in 2020. Shipments would rise to 600 metric tons in 2021, 1,200 metric tons in 2022, and 2,000 metric tons in 2023, and reach the planned maximum annual capacity of 3,000 metric tons in 2024.[2] Because the total U.S. commercial reactor fleet discharges an average of about 2,000 metric tons of spent fuel per year, the above shipment schedule would not begin reducing the backlog of spent fuel stored at reactor sites until 2024.

DOE estimates that the amount of commercial spent fuel stored in pools of water and dry casks at reactor sites and other facilities was 57,700 metric tons at the end of 2007.[3] If commercial spent fuel continues to accumulate at the rate of 2,000 metric tons per year, then inventories would reach 81,000 metric tons before shipments are to begin in 2020 and peak at nearly 85,000 in 2023, after which shipments to Yucca Mountain would exceed reactor discharges by 1,000 tons per year. DOE projections indicate that shipments of all spent fuel from previous and existing U.S. nuclear power plants would continue until about 2066, totaling 109,300 metric tons (under existing reactor license periods and extensions). In addition, the equivalent of 12,800 metric tons of defense-related spent nuclear fuel and high-level radioactive waste would be received at Yucca Mountain during the same period, for a total of 122,100 metric tons.[4]

Although DOE's cost projections assume that all spent fuel from existing reactors, plus defense waste, will be shipped to the planned Yucca Mountain repository, NWPA section 114(d) caps Yucca Mountain's capacity at the equivalent of 70,000 metric tons of spent fuel until a second repository begins operating. No such repository is currently authorized. A recent DOE report on the need for a second repository concludes that all existing and anticipated spent fuel and high- level waste could be physically accommodated at Yucca Mountain.[5] Legislation to lift the 70,000- ton limit proposed by the Bush Administration was introduced during the 109th (H.R. 5360) and 110th (S. 37) Congresses but not acted upon.

As amended in 1987, NWPA provides no backup plan for spent fuel management if the Yucca Mountain repository were to be halted. A "monitored retrievable storage" (MRS) facility is authorized by NWPA section 142, but construction is prohibited until NRC has authorized the construction of the Yucca Mountain repository.

Section 302 of NWPA requires nuclear power plant operators to sign contracts with DOE under which the nuclear plants must pay fees to the federal government in return for DOE's spent fuel disposal services. The nuclear power plant fees are deposited in a Treasury account called the Nuclear Waste Fund to pay for the DOE waste program but cannot be spent without congressional appropriation. The Fund's balance was about $20 billion at the end of FY2008.[6] Because the DOE waste program will also handle defense-related waste, Congress typically supplements annual appropriations from the Nuclear Waste Fund with appropriations from general revenues.

Annual spending for the nuclear waste program, focusing on Yucca Mountain site studies and the license application, has averaged about $400 million in recent years. However, DOE projects that to build the repository and develop a transportation system within the next 12 years, annual funding would need to increase to nearly $2 billion during the peak of construction. Disposal of all 122,100 metric tons of currently anticipated waste at Yucca Mountain is projected to cost $96.18 billion (in 2007 dollars) through 2133.[7]

NWPA section 302(d) restricts the use of the Nuclear Waste Fund to disposal activities and research authorized by the act. The section specifically prohibits DOE from expending the funds for any facility besides those expressly authorized by NWPA or subsequent act of Congress. As amended in 1987, NWPA currently authorizes only a repository at Yucca Mountain and a monitored retrievable storage facility tied to the operation of a Yucca Mountain repository.

The contracts that DOE signed with nuclear utilities required DOE to begin taking waste from reactor sites by January 31, 1998. Because that deadline was missed, DOE has been ruled liable for all waste storage costs that nuclear utilities would not have incurred had shipments to the planned repository begun on time. The U.S. Court of Federal Claims has already issued several judgments against DOE. Claims are paid from the federal judgment fund, rather than the Nuclear Waste Fund, and require no congressional appropriations. DOE calculates that its nuclear waste liabilities to nuclear reactor operators under current law will ultimately total $11 billion if shipments begin by 2020 as currently planned and potentially much more if waste operations are further delayed.[8]

OPTIONS FOR HALTING OR DELAYING YUCCA MOUNTAIN

The Yucca Mountain repository is now in the final stages of the lengthy approval process established by NWPA. Therefore, if the Obama Administration does "not believe that Yucca Mountain is a suitable site," what options are available at this point to stop the project?

NWPA sections 113-116 prescribed the following actions by the Secretary of Energy, the President, and the Nuclear Regulatory Commission toward developing a nuclear waste repository at the Yucca Mountain site:

- The Secretary were to determine whether the site is suitable for a repository, and, if so, notify the State of Nevada and recommend that the President approve the project.
- If the President agreed with the Secretary's recommendation, the President were to submit an approval recommendation to Congress.
- After the presidential recommendation, the Nevada Legislature or Governor were allowed to submit a notice of disapproval to Congress.
- State disapproval would block the repository unless Congress voted within 90 days for an approval resolution that was signed by the President.
- Once the presidential site designation took effect, the Secretary were required to submit a repository license application to NRC.

All these steps have now taken place, and the Yucca Mountain license application has been docketed by NRC for consideration. NRC is an independent regulatory body not directly under the President's control. But the President has a variety of tools at his disposal that could dramatically affect the nuclear waste program.

Withdraw License Application

To stop further action on Yucca Mountain, perhaps the most dramatic step would be for DOE to withdraw the repository license application, as allowed by NRC procedures.[9] Such a withdrawal could be temporary, pending completion of some of the study options described below, or it could be permanent, with the intention of completely ending the Yucca Mountain project. If the license application were permanently withdrawn, the previous presidential site designation may have to be reversed as well, because NWPA section 114 requires that the license application be submitted 90 days after the presidential designation takes effect (a deadline that was missed by more than five years). The fact that, as noted above, the presidential designation of the Yucca Mountain site took effect pursuant to a congressional override of Nevada's "state veto" could be a complicating factor.

Reduce Appropriations

Restricting funding for the Yucca Mountain Project could be another approach. Congressional opponents of the waste program have succeeded in cutting its funds in recent years, although not enough to prevent DOE from submitting the license application. The Bush Administration requested an appropriation of $37.3 million from the Nuclear Waste Fund for NRC's license review activities during FY2009,[10] and substantial reductions could force significant delays in the planned four-year licensing schedule.

As noted above, DOE contends that it will need large funding increases to design and build the Yucca Mountain repository by 2020, so even steady funding would probably push back that schedule by many years. Of course, eliminating funding altogether for NRC licensing and DOE repository construction would halt further work on the project. The Obama Administration's ability to reduce or eliminate Yucca Mountain funding, if it so desired, would depend on its influence with Congress.

Key Policy Appointments

Although NRC is independent of the Administration, and Commissioners cannot be removed by the President without cause, the President can change the makeup of the Commission over time. Significant changes in the Commission could affect the Yucca Mountain licensing process and its ultimate outcome. The five NRC Commissioners serve five-year terms that are staggered so that one expires each year, on June 30. No more than three Commissioners may be from the same political party. One of the five slots is currently vacant. The President can also redesignate the Chairmanship of the NRC to a different Commissioner at any time.

Other presidential appointments could also have a strong effect on the Yucca Mountain project, the Secretary of Energy in particular. Senate Majority Leader Harry Reid had promised to block any nominee for that post who supported the Yucca Mountain site.[11] The Energy Secretary could initiate a redirection of the nuclear waste program, revisit DOE's complex computer models that predict low radioactive releases from Yucca Mountain, or, as noted above, withdraw the license application.

Senator Reid expressed support for President Obama's Energy Secretary, former Lawrence Berkeley National Laboratory Director Steven Chu, contending that "Dr. Chu also knows, like most Nevadans, that Yucca Mountain is not a viable solution for dumping and dealing with nuclear waste."[12] In a 2005 interview posted on the Lawrence Berkeley National Laboratory website, Chu noted projections from his lab that waste canisters in Yucca Mountain would begin to fail after about 5,000 years, which would require the underlying rock formations to prevent unacceptable migration of radioactive material into the groundwater.[13] However, Chu also signed a nuclear policy statement with other national laboratory directors in August 2008 that called for "licensing of the Yucca Mountain Repository as a long-term resource."[14]

Another important appointment for Yucca Mountain is the Administrator of the Environmental Protection Agency (EPA), Lisa Jackson, who was confirmed January 22, 2009. Under NWPA, EPA sets the radiation protection standard that NRC must use in licensing Yucca Mountain. After an earlier version of the standard was struck down by a federal court, EPA published a final standard October 15, 2008, which sets individual radiation exposure limits of 15 millirems for the first 10,000 years after disposal and 100 millirems after 10,000 through one million years.[15] The State of Nevada has sued to overturn the EPA regulations, contending that there is no justification for a higher limit after 10,000 years.[16] If the EPA standard is overturned, licensing of the Yucca Mountain site could become more difficult.[17]

Waste Program Review

Rather than moving immediately to halt the Yucca Mountain repository, the Obama Administration could delay or suspend the project through the approaches described above and then initiate a major program review – a step implied by the Obama-Biden campaign policy statement. The scope of the review could include such topics as management issues, research and development needs, foreign waste management experience, and broad policy options.

Such a review could be conducted by an interagency task force within the new Administration or by an outside entity such as the National Academy of Sciences (NAS) or a presidential commission. Independent scientific reviews of the waste program are currently provided by the Nuclear Waste Technical Review Board (NWTRB), which, under NWPA Title V, the President appoints from nominees provided by NAS. However, the President does not set the agenda for independent agencies such as NWTRB, so it is not clear what role the Board might play in a major policy review.

Consequences of a Yucca Mountain Policy Shift

A decision by the incoming Administration to halt or delay development of the Yucca Mountain repository could have significant impact on the federal budget, proposed new U.S. nuclear power plants, waste storage at existing reactor sites, and disposal of defense-related nuclear waste. Such consequences could be most pronounced if such a policy shift involved simply a halt of Yucca Mountain without legislation to forge a new direction – legislation that presumably would address the issues discussed below.

No matter what decision is made on Yucca Mountain, there is a broad scientific consensus that long-term isolation of nuclear waste from the environment – for at least 1,000 years – will still be required. In other words, if Yucca Mountain were abandoned, another repository site in the United States would almost certainly have to be found eventually. Reprocessing and recycling of nuclear spent fuel can reduce the amount of long-lived radioactive waste requiring isolation but cannot entirely eliminate the need for such isolation. Alternatives to deep geologic waste isolation have been studied, such as space and subseabed disposal, but they face daunting technical obstacles, and none has ever been developed beyond the conceptual stage.

If development of the Yucca Mountain repository were delayed or halted, commercial spent fuel and defense-related nuclear waste would almost certainly remain at numerous on-site storage facilities longer than currently planned. A new repository to replace Yucca Mountain would be unlikely to open by 2020 to prevent delays in DOE's current shipping schedule. Federal centralized interim storage has been proposed repeatedly as a solution, including the MRS facility authorized by NWPA, but no such facility has been developed. If new legislation were to authorize a central interim storage facility, it possibly could begin receiving waste by 2020 and prevent further delays. Another possibility is a private spent fuel storage facility in Utah that has already received an NRC license and might be opened relatively quickly if other administrative approvals were granted (as discussed in a subsequent section). However, the Utah facility's licensed capacity is limited to 40,000 metric tons of spent fuel, which could be stored at the site for no longer than 40 years.

Federal Liabilities for Disposal Delays

As noted above, DOE is liable for utilities' nuclear waste storage costs resulting from the missed NWPA disposal deadline. According to DOE, "for each additional year of delay, the Department estimates that there may be hundreds of millions of dollars of additional damages."[18] These mandatory payments would be a direct cost to the federal government, but they would stretch over several decades because utilities cannot recover damages until their extra storage costs are actually incurred. DOE projects that if disposal begins by 2020, $11 billion in liabilities will be incurred by 2056.[19]

If the Yucca Mountain site were abandoned without an alternative storage or disposal process in place, court judgments against DOE could rise far higher. The nuclear industry has raised the possibility that DOE could be found in complete default on its NWPA contracts and be ordered to refund all the nuclear waste fees that had been collected, in addition to paying utilities' extra at- reactor storage costs.[20] Through the end of FY2008, DOE had collected more than $28 billion in fees and interest payments – an amount that has been growing at about $1.5 billion per year.[21] In at least one of the nuclear utility cases before the Federal Court of Claims, a judge issued a show- cause order for why the DOE nuclear waste contracts should not be voided and all payments returned to utilities,[22] although that step was not included in the court's final decision.[23]

Licensing Complications for New Power Reactors

No new commercial reactors have been ordered in the United States since the 1970s, but concerns over potential carbon dioxide controls and high natural gas prices have prompted U.S. electric utilities to again consider the nuclear power option. License applications for 26 new reactors have been filed with NRC, and more are anticipated.[24] Further delays in the DOE nuclear waste program could pose an obstacle to licensing the proposed new reactors.

NRC established a policy in 1977 that it "would not continue to license reactors if it did not have reasonable confidence that the wastes can and will in due course be disposed of safely."[25] NRC then began a Waste Confidence proceeding that resulted in 1984 findings that there was "reasonable assurance" that a nuclear waste repository would be available by 2007-2009 and that waste could be safely stored at reactor sites for at least 30 years after reactors have shut down.[26]

After DOE's schedule for opening the Yucca Mountain repository slipped to 2010, NRC revised its Waste Confidence Decision in 1990 to find reasonable assurance that a repository "will be available within the first quarter of the twenty-first century."[27] With DOE now planning to open Yucca Mountain by 2020 at the earliest, NRC is proposing a further revision to find reasonable assurance that a repository will be available within 50-60 years after a reactor's licensed operating life and that spent fuel can be stored safely for at least 60 years after a reactor's licensed life.[28] Although the NRC's latest proposed revision would allow for decades of further slippage in the Yucca Mountain schedule, it is not clear that NRC's waste-related criteria for licensing new reactors would be satisfied if the Yucca Mountain project were canceled without an alternative plan in place.

Six states – California, Connecticut, Kentucky, New Jersey, West Virginia, and Wisconsin – have specific laws that link approval for new nuclear power plants to adequate waste disposal capacity. Kansas forbids cost recovery for "excess" nuclear power capacity if no "technology or means for disposal of high-level nuclear waste" is available.[29] The U.S. Supreme Court has held that state authority over nuclear power plant construction is limited to economic considerations rather than safety, which is solely under NRC jurisdiction.[30] No nuclear plants have been ordered since the various state restrictions were enacted, so their ability to meet the Supreme Court's criteria has yet to be tested.

The nuclear waste issue has also historically been a focal point for public opposition to nuclear power. Proposed new reactors that have no clear path for removing waste from their sites could face intensified public scrutiny, particularly at proposed sites that do not already have operating reactors.

Environmental Cleanup Penalties

For defense-related nuclear waste, which resulted from production of nuclear weapons and naval reactor fuel by DOE and its predecessor agencies, indefinite delays in developing a repository could also have legal consequences for DOE. Defense-related high-level radioactive waste resulted from decades of reprocessing spent fuel to extract plutonium for nuclear warheads or highly enriched uranium from spent naval reactor fuel. As noted above, DOE's inventory of defense high-level waste and unreprocessed spent fuel, plus waste from other DOE nuclear programs, totals the equivalent of 12,800 metric tons (the mass of spent fuel before reprocessing). This material is located primarily at Hanford, WA, Savannah River, SC, and the Idaho National Laboratory.

Congress has given states the authority to enforce waste management laws against federal agencies, including DOE. Without Yucca Mountain or an alternative repository plan, DOE would not have a permanent disposal site for waste now stored at its defense-related facilities. A lack of repository capacity, according to DOE, "could threaten the Department's ability to fulfill [regulatory] agreements with the states hosting those sites to remove the waste for permanent disposal."[31]

Long-Term Risk

The near-term environmental impact of further Yucca Mountain delays or abandonment would be minimal for at least 100 years, according to DOE's supplemental environmental impact statement (SEIS) for the repository program.[32] That assessment is consistent with the NRC Waste Confidence Decision cited above. As long as storage facilities at reactors and other sites are maintained and guarded through institutional controls, radioactive releases to the environment are expected to be small.

On-site storage of spent fuel will continue for many decades even if Yucca Mountain begins receiving waste shipments by 2020, as discussed in the baseline program section above. Moreover, some freshly discharged spent fuel will be stored on site as long as reactors are operating. Some environmental groups have argued that it would be safer to leave all nuclear waste for an extended period in hardened on-site storage facilities rather than begin sending it to a central facility as soon as possible, because allowing the waste's radioactivity to decay would reduce the consequences of transportation accidents or sabotage when the waste is ultimately moved.[33] Moreover, waste placed at a central interim storage facility would probably have to be moved a second time to a permanent repository, potentially further increasing transportation risks.

Beyond 100 years or so, the environmental risks of surface storage resulting from the lack of an underground repository become more uncertain. At some point in the future, maintenance and security of surface storage facilities would be expected to drop below adequate levels because of unforeseen circumstances. Whether that would occur after 500 years, 1,000 years, or 10,000 years is open to speculation. Federal nuclear waste policy would presumably continue to envision a permanent disposal method well before such a breakdown, but the risk would be expected to rise if delays continued indefinitely. The SEIS predicts that substantial amounts of radioactivity would reach the accessible environment within 10,000 years after the end of institutional controls on surface storage facilities, "with eventual catastrophic consequences for human health."[34]

Nuclear Waste Policy Options

Because NWPA specifies that only Yucca Mountain may be considered for a repository site and that a federal storage facility cannot open before the repository is licensed, the government's waste management options are sharply limited under current law. Without congressional action, alternatives to Yucca Mountain would consist primarily of indefinite on-site storage or licensing of new private storage sites.

New legislation would open up much broader possibilities, ranging from a search for a new repository site and federal interim storage to reprocessing and alternative disposal technologies. Some – but probably not all – of the consequences of changing the current waste policy could also be mitigated through legislation. Any legislation dealing with nuclear waste siting is almost certain to prove extraordinarily controversial.

Institutional Changes

Almost since the beginning of the current nuclear waste program in 1982, DOE has regularly been accused of mismanagement and allowing political considerations to affect scientific decisions. Many proposals have been made by the nuclear industry and its critics alike to transfer the nuclear waste program to an independent organization that might be more efficient and less affected by politics. For example, a proposal in the 110th Congress (H.R. 6001, section 186) would have handed the DOE waste program to an independent High Level Waste Authority, headed by a presidentially appointed seven-member board. In implementing the nuclear waste program, the Waste Authority would have been authorized to consider all reasonable options, including alternative repository locations.

An independent waste agency could be a government agency, as in H.R. 6001, a government corporation, or a private-sector entity. Two studies conducted for DOE's Global Nuclear Energy Partnership recommended the establishment of a government corporation to manage nuclear waste, including spent fuel reprocessing. One of the studies, by a consortium led by the French firm Areva, called a government corporation the best option "because it allows the utilities to have some level of oversight while full government ownership keeps the cost of capital low."[35] A team led by EnergySolutions recommended that the government corporation have a board of directors drawn from the nuclear industry, with an independent oversight board "to assure that it meets its charter obligations."[36]

Licensing and regulation by NRC could be continued unchanged under any option. The existing funding system could also be transferred largely unchanged to a new government agency, but a new funding system would probably be needed for a private entity. Whether a private entity should take permanent title to all nuclear waste could also be an issue. The nuclear industry has long contended that nuclear waste payments by reactor owners should be available directly for nuclear waste disposal activities without the need for congressional appropriation. But even if the program's management were improved, it is far from clear whether any new waste management organization could avoid the political controversy that has persistently accompanied the DOE program.

Short of establishing an entirely new organization to run the waste program, it has been suggested that additional independent oversight could improve public confidence in DOE's decisions. This was a major reason cited for creating the Nuclear Waste Technical Review Board in 1987. Some have proposed that technical oversight by a non-federal agency would be more credible, such as the Environmental Evaluation Group (EEG) in New Mexico.[37] Congress in 1988 required DOE to sign a contract with a New Mexico university, the New Mexico Institute of Mining and Technology, to administer the EEG to provide independent review and evaluation of the Waste Isolation Pilot Plant (WIPP), a DOE repository for relatively low-radioactivity defense waste (P.L. 100-456, section 1433). The EEG closed in 2004 after DOE halted its funding,[38] and oversight activities are now carried out by the New Mexico Environment Department.[39]

Extended On-Site Storage

It appears unlikely, based on the history of the nuclear waste program, that any alternative storage or disposal sites could become operational earlier than the planned opening of Yucca Mountain in 2020. Therefore, any alternative to the Yucca Mountain repository would almost certainly result in longer on-site storage of nuclear waste than under the current baseline program. Essentially, extended on-site storage is the default option, with the only question being how long. On-site storage could be extended for decades under some policy changes, such as a restart of the repository site search or the pursuit of alternative disposal technologies. Waste might be moved more quickly if central interim surface storage facilities were developed (as discussed below), but designating such sites may be nearly as controversial as siting a repository.

On-site nuclear waste storage at reactor facilities takes place primarily in deep pools of water that are built into the reactor building. The water is necessary to provide cooling and radiation shielding for extremely radioactive spent fuel that is freshly discharged from the reactor. After its radioactivity has sufficiently decayed, usually after several years, spent fuel can be transferred to dry storage casks and stored outside the pools. Most spent fuel pools were not designed to hold all the spent fuel generated during a reactor's operating life. Therefore, as DOE's target date for taking spent fuel from reactor sites has slipped, nuclear reactor operators have had to expand their on-site dry storage capacity (and have sued DOE for compensation).

NRC considers extended on-site storage to be safe as long as storage facilities are adequately maintained and guarded, as discussed above. However, the National Academy of Sciences (NAS) determined in 2005 that spent fuel pools could be vulnerable to terrorist attacks, particularly as spent fuel has been stored more densely in the pools to increase their capacity. NAS found that an attack could drain the cooling water from a spent fuel pool and cause the spent fuel's zirconium cladding to overheat and catch fire, releasing "large quantities of radioactive materials to the environment."[40] The Energy and Water Development Appropriations Act for FY2006 (P.L. 109- 103) included $21 million for NRC to assess the vulnerabilities found by NAS at each reactor site.

Keeping nuclear waste at reactor sites and other existing facilities has long been a major goal of Yucca Mountain opponents. Numerous bills have been introduced over the years to require DOE to take over all responsibility for storing spent fuel at reactor sites, including ownership of the waste and on-site storage facilities (such as S. 784 in the 110th Congress). Such an on-site DOE takeover is intended to reduce utilities' costs, and resulting federal liabilities, for DOE's failure to remove the spent fuel. However, states with nuclear reactors oppose indefinite on-site storage, and utilities and state regulators have opposed using the Nuclear Waste Fund to pay for on-site storage (as proposed by S. 784) rather than permanent disposal or central interim storage.

Federal Central Interim Storage

DOE does not believe it has the authority under current law to develop a central interim nuclear waste storage facility other than the "monitored retrievable storage" facility authorized by NWPA,[41] and the 1987 NWPA Amendments prohibit such an MRS facility from opening until Yucca Mountain is licensed. Moreover, construction of an MRS cannot begin until NRC grants a construction permit for the repository; the MRS is limited to 15,000 metric tons of spent fuel; and the MRS cannot be located in Nevada (NWPA sections 145 and 148). Numerous legislative efforts have been mounted since the mid-1990s to establish central interim storage capacity without the MRS restrictions, but without success. No matter how they have been structured, such proposals have consistently faced overwhelming concerns that any "interim" storage facility would undercut political support for a permanent repository and therefore become a "de facto" permanent disposal site.

Central interim nuclear waste storage facilities would use dry cask technology that is similar to that currently in place at reactor sites. Sealed waste canisters would be placed in individual above-ground concrete casks or bunkers for radiation shielding and for cooling by natural air circulation. Because the systems are modular, they can be constructed relatively quickly.

And because the waste is not expected to remain in storage after active maintenance of the facility has ceased, a wide variety of sites are likely to be considered geologically suitable. However, public concern about large quantities of highly radioactive waste stored for an extended period of time, and perhaps indefinitely, has proven to be a major obstacle to central storage proposals.

DOE had proposed to build an MRS facility near Oak Ridge, TN, as a central receiving point for small waste shipments from individual plant sites east of the Rocky Mountains. Spent fuel was to be repackaged if necessary at the MRS facility for consolidated long-distance shipments to the planned Western repository.[42] The Oak Ridge selection was specifically nullified by the 1987 NWPA amendments, which established a new siting procedure along with the restrictions listed above. The 1987 Amendments also established an alternative, voluntary siting procedure for the MRS and other nuclear waste facilities. Under Title IV, a presidentially appointed "nuclear waste negotiator" was authorized to reach agreements with any states or Indian tribes to host nuclear waste facilities under any "reasonable and appropriate" terms. Such agreements could not take effect without being enacted into law, however.

By the early 1990s, finding a voluntary site for a central storage facility appeared to be DOE's best chance for meeting NWPA's 1998 waste acceptance deadline. DOE began providing feasibility study grants to potential volunteers, mostly Indian tribes. Potential agreements with Indian tribes proved highly objectionable to the states in which the tribes were located, and Congress blocked the grant funding in October 1993 (P.L. 103-126). The authority for the nuclear waste negotiator expired on January 21, 1995, without any proposed siting agreements having been reached.

The next major legislative push for an alternative to the MRS took place after it became clear that DOE would be unable to meet the NWPA disposal deadline. In the 104th Congress, nuclear power supporters developed legislation to authorize DOE to open an interim surface storage facility at the Yucca Mountain site by 1998, well before the then-anticipated opening of the underground repository in 2010 (H.R. 1020). The State of Nevada and nuclear power opponents contended that waste should not be transported to Yucca Mountain before the repository was licensed, because if the repository ultimately did not receive a license, the waste would have to be moved again, increasing potential transportation risks and costs. The bill was approved by the Commerce Committee (H.Rept. 104-254) but was not enacted. Similar bills were passed by the House and Senate in the 1 05th Congress (H.R. 1270, S. 104), but President Clinton threatened a veto and a conference was not held.

A final try in the 106th Congress (S. 1287) drew a presidential veto that was narrowly sustained in the Senate.

After the proposals to develop interim storage capacity at Yucca Mountain were rejected, language was included in several appropriations bills and reports to require DOE to store commercial spent fuel at unspecified federal sites. The House Appropriations Committee included language in its report on the FY2006 Energy and Water Development Appropriations Bill to require DOE "to begin the movement of spent fuel to centralized interim storage at one or more DOE sites within fiscal year 2006" (H.Rept. 109-86), although the Senate did not go along with the idea. The Senate Appropriations Committee included an extensive provision in its version of the FY2007 Energy and Water bill (H.R. 5427, section 313) to authorize the Secretary of Energy to designate interim storage sites for spent nuclear fuel. The proposal, which was not enacted, would have required the Secretary to designate a storage site in each state with a nuclear power plant, after consultation with the governor, or to designate regional storage facilities.

A more limited central storage proposal was aimed solely at nine decommissioned reactor sites. Because the decommissioned sites have no ongoing nuclear activities except spent fuel storage, the removal of spent fuel would allow all nuclear-related maintenance and security at those locations to cease, producing significant operational cost savings, according to the nuclear industry. The House Appropriations Committee included report language with the FY2008 Energy and Water bill requiring DOE to "develop a plan to take custody of spent fuel currently stored at decommissioned reactor sites" (H.Rept. 110-185). The resulting DOE report concluded that all 2,800 metric tons of spent fuel at the nine decommissioned sites could be shipped to a federal central storage facility by 2018, but that DOE had no statutory authority to implement such a plan.[43]

DOE has taken spent fuel for storage at its facilities in the past in special cases, such as the damaged core from the 1979 Three Mile Island accident and from the unique Fort Saint Vrain gas-cooled reactor in Colorado. The Three Mile Island core material was shipped to DOE's Idaho National Laboratory for research. DOE is storing the Fort Saint Vrain spent fuel pursuant to a cooperative agreement signed with the reactor supplier and local utility before the demonstration reactor was built.[44] In addition, DOE stores highly enriched, U.S.-origin spent fuel from foreign research reactors because of its potential use in nuclear weapons. Some have contended that those precedents indicate that DOE has sufficient general authority under the Atomic Energy Act to store larger amounts of commercial spent fuel. However, DOE contends that its broad authority under the Atomic Energy Act is restricted to narrow circumstances under the more recently enacted and specific waste management provisions of NWPA.[45]

Private Central Storage

Although DOE does not believe it has authority under current law to construct a federal central interim storage facility for commercial nuclear waste, NRC regularly licenses private-sector interim storage facilities under the Atomic Energy Act.[46] Such "independent spent fuel storage installations" typically are licensed for on-site storage at reactor sites, but they can also include central storage facilities.

After a nearly nine-year licensing process, NRC issued a license for a private central storage facility on February 21, 2006, that was intended to receive waste from commercial reactor sites.[47] The facility was to be developed by a nuclear utility consortium called Private Fuel Storage (PFS) on the reservation of the Skull Valley Band of the Goshute Indians in Utah. The 20-year license, renewable for an additional 20 years, allows up to 40,000 metric tons of spent fuel to be stored in 4,000 dry casks pending shipment by DOE to a permanent repository. PFS will not take title to the spent fuel, so waste is to be returned to the utilities that own it if DOE cannot take it away before the PFS license expires.[48]

On September 7, 2006, the Department of the Interior issued two decisions blocking the PFS project. The Bureau of Indian Affairs disapproved a proposed lease of tribal trust lands to PFS, concluding there was too much risk that the waste could remain at the site indefinitely, among other objections.[49] The Bureau of Land Management rejected the necessary rights-of-way to transport waste to the facility, concluding that a proposed rail line would be incompatible with the Cedar Mountain Wilderness Area and that existing roads would be inadequate.[50] Contending that the Interior Department was motivated by political pressure from the State of Utah, which strongly opposed the facility, the Skull Valley Band of Goshutes and PFS filed a federal lawsuit July 17, 2007, to overturn the decisions.[51]

The PFS project was intended to provide a waste storage option for nuclear plants that might have trouble gaining approval for on-site storage facilities or decommissioned reactors that want to remove remaining spent fuel from their sites. If the PFS facility were considered potentially useful as part of a revised spent fuel strategy, the new Administration could revisit the Interior

Department's administrative decisions that are blocking the project. However, if those decisions were reversed, the project would still need to overcome a challenge to the NRC license filed by the State of Utah.[52] Another consideration for this option is that, because the waste would have to be returned after 40 years, and utilities would be paying for the service, the PFS facility might not significantly reduce DOE's liabilities for delays in spent fuel acceptance.

Spent Fuel Reprocessing and Recycling

The major alternative to direct disposal of spent fuel (the "once through" fuel cycle) is the "closed" fuel cycle, in which spent fuel is reprocessed into new fuel. The closed fuel cycle could reduce the volume and long-term radioactivity of nuclear waste and potentially postpone the need for permanent disposal. However, a National Academy of Sciences study of reprocessing technologies found that "none of the S&T [separations and transmutation] system concepts reviewed eliminates the need for a geologic repository."[53] Recycling spent fuel could also greatly increase the amount of energy extracted from a given supply of uranium. However, the closed fuel cycle is generally considered to be substantially more expensive than the once-through cycle.[54] Moreover, the separation of plutonium from spent fuel has long been a subject of national policy debates because of its potential role in nuclear weapons proliferation.

Fuel for U.S. nuclear reactors currently consists of uranium in which the fissile isotope U-235 has been increased (enriched) to 3-5%, with the remainder being the non-fissile isotope U-238. During the fuel's several-year irradiation period in the reactor, most of the U-235 splits, or fissions, releasing energy. Some of the U-238 is transmuted into fissile isotopes of plutonium, some of which also fissions. In reprocessing, the uranium and plutonium are chemically separated to be made into new fuel, while the lighter elements resulting from the fission process, called fission products, are stored for disposal.

New fuel made from reprocessed uranium and plutonium can be recycled in existing commercial light water reactors, which is being done in other countries, primarily France. After being recycled once, however, the buildup of undesirable plutonium isotopes makes further recycling in today's commercial reactors problematic. Without multiple recycling, the plutonium and other long-lived isotopes cannot be fully fissioned or transmuted into shorter-lived radioactive isotopes, and the benefits for waste disposal would therefore be modest.

For multiple recycling of spent fuel, advanced reactors would be necessary. DOE has evaluated a wide variety of options as part of its Global Nuclear Energy Partnership (GNEP) program. These include initial recycling in existing light and heavy water reactors, and subsequent recycling in high-burnup gas-cooled reactors, reactors fueled by thorium and plutonium, and "fast" reactors (in which neutrons are not slowed by water or other materials).[55] A reprocessing and recycling system with sufficient capacity could eventually treat existing spent fuel inventories along with newly generated spent fuel.

For waste disposal, the goal of such a recycling system would be to send only the fission products and other short-lived radioisotopes to a permanent repository and feed all the uranium, plutonium, and other long-lived radioisotopes back into a reactor after each cycle. If that could be accomplished, the nuclear waste in a repository would decay to insignificant levels within about 1,000 years and eliminate longer-term uncertainty about the repository's performance. Spent fuel recycling could also save space in an underground repository by reducing the near-term heat load, which is the primary limit on repository capacity.[56] To address nuclear nonproliferation concerns, the GNEP program is conducting research on reprocessing technology that would not separate plutonium in a pure enough form for direct use in nuclear weapons.

A potential nearer-term benefit of a reprocessing strategy would be to provide an alternative destination for spent fuel currently stored at reactor sites if Yucca Mountain were to be abandoned. Because DOE is still conducting R&D on a variety of possible reprocessing technologies, however, a U.S. reprocessing facility would probably not open earlier than the current Yucca Mountain target date.

Earlier waste shipments might be possible under proposals that have been made for foreign reprocessing of U.S. spent fuel. Several foreign reprocessing plants are currently in operation. The Senate Energy and Natural Resources Committee included a provision in a nuclear waste bill in the 104th Congress (S. 1271) that would have authorized DOE to take title to spent fuel and ship it to a reprocessing plant in England. However, the provision proved highly contentious and was dropped from the final bill passed by the Senate (S. 1936). In a 2008 report for GNEP, a consortium led by the French nuclear firm Areva recommended that U.S. spent fuel be reprocessed overseas from 2010 to 2019 before the startup of a U.S. reprocessing plant after 2020.[57] The use of inactive defense-related reprocessing facilities at DOE's Savannah River Site in South Carolina has also been suggested for U.S. commercial spent fuel.[58]

The amount of spent fuel that could be shipped to reprocessing plants would be another consideration. Existing reprocessing plants in France and England are designed to handle about 800 metric tons of spent fuel per year. Therefore, at least three plants of that size would need to be constructed in the United States (assuming minimal foreign reprocessing) to handle the 2,000 metric tons of spent fuel discharged annually from U.S. reactors. About four plants would be needed to exceed the planned shipment rate to Yucca Mountain.

Many decades would be required to implement a reprocessing and recycling strategy. For example, the Areva consortium projected that a steady-state recycling system would not be fully in place until about 2070, even if the currently planned 63,000 metric tons of spent fuel were emplaced in Yucca Mountain rather than being reprocessed. The first U.S. reprocessing plant would become operational after 2020.[59] Proposals by three other GNEP consortia included similar time frames.

Non-Repository Options

The inherent difficulty of siting a permanent geologic repository for high-level nuclear waste has led to a variety of proposals over the past few decades for non-repository disposal options. NWPA section 222 authorizes DOE to conduct research on such disposal alternatives. The most seriously analyzed ideas involve launching waste into space or burying it in the deep seabed. Some plausible concepts for implementing these ideas have been developed, but a great deal of development work would still be required to determine their likely feasibility.

Congress established a DOE Office of Subseabed Disposal Research in the Nuclear Waste Policy Amendments Act of 1987 (P.L. 100-203). The office was required to organize a Subseabed Consortium among leading research institutions to develop a research plan for identifying subseabed disposal sites, developing conceptual designs for subseabed disposal systems, and assessing potential environmental impacts. However, few resources were provided for the subseabed office before it was abolished in 1996 by P.L. 104-66.[60]

Previous research on subseabed disposal was conducted by the Nuclear Energy Agency (NEA) of the Organization for Economic Cooperation and Development. The United States participated in the effort through the DOE

Subseabed Disposal Project, on which about $125 million was spent from 1974-1986.[61] The NEA program studied the emplacement of nuclear waste canisters in ocean sediments with gravity-driven penetrators or in drilled holes. NEA concluded in 1988 that the sediments would probably contain the waste well enough to keep the maximum dose to humans – occurring after about 100,000 years – "many orders of magnitude below present standards" and pose "insignificant risk to the deep sea environment." However, NEA also concluded that more research would be needed to confirm the safety of the subseabed disposal concepts.[62]

Subseabed disposal is currently prohibited under the 1996 Protocol to the 1972 London Dumping Convention, which was signed by the United States on March 31, 1998, and entered into force March 24, 2006, but has not been ratified by the Senate. The Protocol amended the definition of "dumping" to include "any storage of wastes or other matter in the seabed and the subsoil thereof from vessels, aircraft, platforms or other man-made structures at sea."[63] Previously it had been unclear whether the Convention prohibited subseabed disposal. Annex 1 of the Protocol requires parties to the agreement to complete a scientific study of sea disposal of radioactive material other than high-level waste by 2019 and every 25 years thereafter.

Disposal of nuclear waste in outer space has also been studied by DOE and its predecessor agencies. In a 1974 draft environmental statement on nuclear waste management, the Atomic Energy Commission (AEC) reviewed government studies of such concepts as "solar system escape, solar impact, high-earth orbit, and a solar orbit other than that of the planets." The report concluded that space disposal "does not seem an attractive alternative to the geological development program."[64] Major concerns include launch costs, launch safety, and the potential for future waste re-entry into Earth's atmosphere. Proposed alternatives to conventional rocket- based launch systems, such as laser propulsion and electromagnetic rail guns, might have safety and cost advantages, but a major federal commitment would be needed to determine their feasibility.[65]

Other disposal concepts studied by DOE and its predecessors include waste emplacement in polar ice sheets, deep boreholes, and deep well injection of liquid waste. AEC's draft environmental statement dismissed those alternatives as not viable,[66] and they have since received relatively little attention.

New Repository Site

Even if nuclear waste is placed in extended surface storage and is reprocessed to remove the longest-lived radioactive isotopes, a permanent disposal method almost certainly would still be required. Barring the non-repository options discussed above, that would mean that the abandonment of Yucca Mountain for any reason would eventually require a search for another repository site.

The history of site selection efforts under NWPA indicates that a new repository site search would be slow-moving and extremely controversial. Vast areas of the United States would again be under consideration, after having been eliminated by the 1987 congressional designation of Yucca Mountain as the sole candidate site. Every decision made by whatever entity were to be placed in charge of the site search would probably face intense opposition, especially as the search began to narrow. Designing a selection process that could overcome such pressures would be a major challenge.

NWPA was intended to set up a fair and technically sound process for selecting among numerous potential repository sites that DOE and its predecessor agencies had been considering. Without such a legislative mandate, DOE's previous efforts to find a waste site appeared unlikely to overcome the controversy that had arisen at every potential location.[67] However, the explicit waste siting process created by NWPA lasted only about five years before being paralyzed by renewed controversy.

Under NWPA as originally enacted, the Secretary of Energy was required to establish guidelines that DOE would follow in nominating at least five suitable sites, of which three were to be recommended to the President for detailed study, or "characterization" by January 1, 1985. Sites that DOE had been considering included salt domes along the Gulf Coast, bedded salt in the Great Plains and Midwest, volcanic tuff in the West, and basalt in the Pacific Northwest. Energy Secretary John S. Herrington recommended Hanford, WA; Deaf Smith County, TX, and Yucca Mountain for site characterization on May 27, 1986.[68] After completing the characterization of the three sites, the Secretary was to recommend one of them to the President for the nation's first permanent nuclear waste repository. The President was required to submit his choice to Congress by March 31, 1987.

To address concerns about whether a single site or region should take all of the nation's high-level nuclear waste, NWPA limited the first repository to 70,000 metric tons until a second repository was opened. A separate track was established for locating a second repository site. By July 1, 1989, the Secretary of Energy was to nominate five sites for a second repository, including at least three sites that had not been among the five sites nominated for the first repository, and recommend three of them to the President. The recommended sites were to be located, "to the extent practicable," in different geologic media. The President was to recommend a second repository site to Congress by March 31, 1990, from any of the sites previously characterized.

Unlike the process for the first repository, which started with specific candidate sites that were already under consideration, the site search for the second repository was conducted more systematically. DOE began by focusing on major formations of granite and other crystalline rock, which had not been included in the first repository effort, in 17 states in the upper Midwest and Atlantic coast. In consultation with states, DOE developed a screening methodology to rank candidate bodies of rock for their potential suitability as a repository.[69] DOE released preliminary rankings that identified 12 promising rock bodies in seven states in January 1986.[70]

DOE's identification of potential sites for the second repository drew intense opposition from the affected states. The three potential host states for the first repository also raised strong objections, which intensified when Secretary Herrington announced on May 28, 1986, that work on the second repository would be indefinitely postponed. Herrington said the decision was based on lower growth projections for nuclear power that delayed the need for a second repository, but officials from the first repository candidate states in the West contended that the Reagan Administration had responded to political pressure from the Eastern candidate states and had unraveled a key regional compromise in NWPA.[71] Opposition from Tennessee to DOE's proposed MRS site near Oak Ridge added to the controversy. The Senate Appropriations Committee made note of the deteriorating situation:

> Intense and widespread criticism, controversial programmatic decisions by the Secretary of Energy, and a proliferation of substantial litigation have taken a toll on progress toward the goals of the program.[72]

In addition to the controversy over site selection, it had become apparent that NWPA's timelines for characterizing the candidate sites and the anticipated cost of the characterization effort were unrealistic. With the future of the nuclear waste program in doubt, the 100th Congress decided to reopen NWPA for fundamental revision. The resulting NWPA Amendments Act of 1987 cancelled the second repository program, nullified DOE's selection of Oak Ridge for an MRS facility, and statutorily designated Yucca Mountain as the sole candidate site for a repository. Supporters of the legislation contended that characterizing only one site rather than three would be faster and save money, and noted that Yucca Mountain had been the most highly rated of the three candidates by DOE.[73] Some lawmakers, however, contended that the statutory designation of Yucca Mountain was made primarily for political reasons.[74]

The NWPA Amendments Act also provided for annual payments as an inducement to states for hosting nuclear waste facilities. States could receive up to $20 million per year for hosting a repository and $10 million for an MRS site if they agreed not to exercise their right under the law to disapprove those facilities. However, Nevada expressed no interest in the payments and, as noted previously, exercised its "state veto" in 2002. DOE did not conduct an MRS site search under the NWPA Amendments, relying instead on the Nuclear Waste Negotiator to find a voluntary site, as discussed earlier.

Although naming a single site for characterization was intended to speed up the development of a nuclear waste repository, the process actually took another 15 years after the 1987 Amendments, plus another five years to complete the license application to NRC. Supporters of the waste program contend that chronic underfunding by Congress was a major reason for the slow progress, while opponents primarily blamed DOE management problems. The State of Nevada was also able to slow the repository by denying state permits for various characterization activities and through successful lawsuits, such as the challenge to EPA's environmental standards. Nevada filed a 1,500-page petition with NRC in December 2008 to intervene in the Yucca Mountain license proceeding, raising dozens of safety, environmental, and other contentions.[75]

The history of U.S. efforts to site a nuclear waste repository illustrates the difficulty in successfully addressing local, state, and regional objections to such facilities. The United States did not succeed with the administrative process started by the Atomic Energy Commission, with the site-ranking system used for the NWPA first repository selection, the broad screening process used for the second repository, the benefits offered under the NWPA Amendments, or the voluntary selection process by the Nuclear Waste Negotiator. If Yucca Mountain is abandoned, that would arguably spell failure for the statutory designation method as well.

Just because those approaches were unsuccessful in the past does not mean they could not work in the future with program design modifications, better management, and changed circumstances. For example, it has been recently suggested that a negotiated benefits agreement with Nevada might now be feasible, given the current economic downturn.[76] During the debate on the 1987 Amendments, Representative Morris Udall, Chairman of the House Interior Committee, contended that the original NWPA selection process would have worked had it been implemented properly:

> We created a principled process for finding the safest, most sensible places to bury these dangerous wastes. We were confident that while no State wanted a nuclear waste repository, the States ultimately chosen would accept the outcome because the selection process would have been fair and technically credible.
>
> Today, just 5 years later, this great program is in ruins. To help a few office seekers in the last election, the administration killed the eastern repository program, shattering the delicate regional balance at the heart of the 1982 act. Since then the Western States have felt they are being treated unfairly, and they no longer trust the technical integrity of the Department of Energy's siting decisions.[77]

Others have expressed doubt that a purely scientific and objective selection process is possible, given the inherent difficulties in making extremely long-term projections of repository behavior. The Director of the Office of Civilian Radioactive Waste Management recently described the siting of nuclear waste facilities as a "technically informed political decision."[78]

DOE's long but ultimately successful struggle to open a deep geologic repository for mid-level waste – the WIPP facility near Carlsbad, NM – indicates that siting of nuclear waste facilities is not necessarily impossible. State and local officials had invited AEC to consider the deep salt beds in the economically depressed area in the early 1970s for a high-level waste repository. After a great deal of statewide controversy, although with consistent local support, Congress authorized WIPP in 1979 to hold defense-related transuranic waste (P.L. 96-1 64).[79]

Transuranic (TRU) waste is not considered to be as hazardous as spent fuel and high-level waste, but it nevertheless requires long-term isolation in a geologic repository. TRU waste consists of relatively low-radioactivity material contaminated with more than a minimum concentration of long-lived plutonium.

DOE's efforts to implement the 1979 WIPP authorization were hampered by concerns by state officials that spent fuel and high-level waste would eventually be disposed of along with the transuranic waste.[80] After a dozen years of controversy over the project's implementation, Congress in 1992 enacted the Waste Isolation Pilot Plant Land Withdrawal Act (P.L. 102-579), detailing the regulations and procedures that DOE would have to follow to open the facility and banning high-level waste and spent fuel. Slow progress prompted Congress to amend the WIPP Land Withdrawal Act in 1996 to exempt WIPP waste from some land disposal restrictions and provide $20 million for New Mexico bypass roads for waste shipments (P.L. 104-201). The first waste was shipped to the repository in March 1999, nearly 20 years after the facility was authorized.[81]

It has recently been suggested that WIPP again be considered as a site for high-level waste disposal. Spent nuclear fuel could be more technically problematic, because the physical flow of the salt within a period of years will close in on stored waste and eliminate the option to retrieve the waste after 100 years or so, as could be done at Yucca Mountain.[82] Such "salt creep" occurs more quickly at higher temperatures, which could result from the disposal of high-level waste and spent fuel. A potential advantage of salt creep is that it can provide a natural seal around the waste.[83] Nevertheless, the State of New Mexico continues to strongly oppose any disposal of high-level waste at WIPP.[84]

CONCLUDING DISCUSSION

Significant scientific uncertainty – if not clear technical unsuitability – has arisen at every potential high-level nuclear waste repository site evaluated by the federal government. Such doubts have fed the public controversy that inevitably accompanies the announcement of such sites. As a result, the federal government has not succeeded in opening any central facilities for permanent disposal or interim storage of spent nuclear fuel and high-level waste.

The controversial nature of siting nuclear waste facilities increases the likelihood that alternatives to the proposed Yucca Mountain repository would leave waste at existing storage sites longer than under the current program schedule. Major consequences under current law could include increased liability by the federal government for utility storage costs, and fines and penalties for missing cleanup deadlines at defense-related nuclear facilities. Although NRC has determined that waste can be stored safely at reactor sites for many decades, the licensing of new plants could be affected by the lack of a definite disposal plan. Extremely long disposal delays would also increase the risk that adequate maintenance and security at storage sites would end before the waste could be removed.

Central interim storage of nuclear waste has regularly been suggested as the quickest way to begin moving waste from existing storage sites. However, without a plan for permanent disposal, the development of interim sites could be especially controversial. Reprocessing of spent fuel has long been proposed as a way to reduce the hazards of nuclear waste by removing plutonium and other long-lived radioactive material. While such a technological approach could make it easier to site a permanent repository, the separation of plutonium raises significant opposition because of its potential use in nuclear weapons and effects on U.S. nonproliferation policy. DOE is researching reprocessing techniques that could reduce the separation of pure plutonium, but their effectiveness and potential high cost continues to be a subject of controversy.

The 1987 designation of Yucca Mountain as the nation's sole candidate site for a national high- level nuclear waste repository was a calculated risk that the site could be developed successfully. There is no backup plan in place. Yucca Mountain opponents contend that, as a result, the federal government has stuck with the site no matter what technical problems have been discovered. But if Yucca Mountain is determined to have significant problems, an alternative course will have little existing policy framework to build upon.

End Notes

[1] Obama for America, "Barack Obama and Joe Biden: New Energy for America," campaign issue statement, 2008, http://www.barackobama.com/pdf/factsheet_ energy

[2] U.S. Department of Energy, Office of Civilian Radioactive Waste Management, *Total System Life Cycle Cost Report*, DOE/RW-0591, Washington, DC, July 2008, p. 20, http://www.ocrwm.doe.gov/about/budget/pdf/ TSLCC_2007_8_05_08.pdf.

[3] U.S. Department of Energy, Office of Civilian Radioactive Waste Management, *Report to the President and the Congress by the Secretary of Energy on the Need for a Second Repository*, DOE/RW-0595, Washington, DC, December 2008, p. 6, http://www.rw.doe.gov/info_library/program

[4] DOE, *Life Cycle Cost Report,* op. cit., p. A-3.

[5] DOE, *Need for a Second Repository,* op. cit., p. 1.

[6] Department of Energy, Office of Civilian Radioactive Waste Management, *Monthly Summary of Program Financial & Budget Information*, as of September 1, 2008, p. 7.

[7] U.S. Department of Energy, Office of Civilian Radioactive Waste Management, *Total System Life Cycle Cost Report*, DOE/RW-0591, Washington, DC, July 2008, p. B-2.

[8] Christopher A. Kouts, Office of Civilian Radioactive Waste Management, "Yucca Mountain Program Status Update," Presentation to Environmental Protection Agency Workshop on Energy and Environmental Sustainability in a Carbon Constrained Future, New York, NY,

September 11, 2008, p. 9, http://www.epa.gov/region2/energyworkshop/workshop_presentations/session2/nuclear_session/panel1_nuclear_waste_disposal.pdf.

[9] 10 C.F.R. § 2.107.

[10] Nuclear Regulatory Commission, *Performance Budget Fiscal Year 2009*, NUREG-1100 Volume 24, February 2008, p. 2, http://www.nrc.gov/reading

[11] Erica Werner, "Reid Won't Allow Energy Secretary Who Supports Yucca Waste Dump," *Associated Press*, December 4, 2008, http://www.rgj.com.

[12] Senator Harry Reid, "Statement on the nomination of Dr. Steven Chu to be Secretary of Energy," press release, December 15, 2008, http://reid.senate

[13] Lawrence Berkeley National Laboratory, "Growing energy: Berkeley Lab's Steve Chu on what termite guts have to do with global warming," press release, September 30, 2005, http://berkeley.edu/news/media

[14] DOE National Laboratory Directors, *A Sustainable Energy Future: The Essential Role of Nuclear Energy*, August 2008, p. 1, http://www.ne.doe.gov/pdfFiles/rpt_SustainableEnergyFuture_Aug2008.pdf.

[15] Nuclear Regulatory Commission, "Public Health and Environmental Radiation Protection Standards for Yucca Mountain, Nevada," 73 *Federal Register* 61256, October 15, 2008.

[16] *State of Nevada v. U.S. Environmental Protection Agency* (U.S. Court of Appeals for the District of Columbia Circuit 2008).

[17] For more on this issue, see CRS Report RL34698, *EPA's Final Health and Safety Standard for Yucca Mountain*, by Bonnie C. Gitlin.

[18] Department of Energy, Office of Civilian Radioactive Waste Management, *Report to Congress on the Demonstration of the Interim Storage of Spent Nuclear Fuel from Decommissioned Nuclear Power Reactor Sites*, DOE/RW-0596, Washington, DC, December 2008, p. 6, http://www.rw.doe.gov/info_library/programES_Interim_Storage_Report_ 120108.pdf.

[19] Kouts, op.cit., p. 9.

[20] U.S. Congress, Senate Committee on Energy and Natural Resources, *Nuclear Waste Litigation*, To examine the impacts of federal court decisions on breach of federal nuclear waste contracts, 106th Cong., 2nd sess., September 28, 2000, S.Hrg. 106-918 (Washington: GPO, 2001), p. 46.

[21] Department of Energy, Office of Civilian Radioactive Waste Management, *Monthly Summary of Program Financial & Budget Information*, as of September 1, 2008, p. 7.

[22] Jeff Beattie, "Federal Judge Suggests Voiding Utilities' Yucca Mountain Contracts," *Energy Daily*, April 29, 2005, p. 1.

[23] *Sacramento Municipal Utility District v. United States*, (Court of Federal Claims 2006).

[24] Nuclear Regulatory Commission, "Combined License Applications for New Reactors," http://www.nrc.gov/reactors/ new-reactors/col.html.

[25] Nuclear Regulatory Commission, 42 *Federal Register* 34391, July 5, 1977.

[26] Nuclear Regulatory Commission, 49 *Federal Register* 34658, August 31, 1984.

[27] Nuclear Regulatory Commission, 55 *Federal Register* 38472, September 18, 1990.

[28] Nuclear Regulatory Commission, "Waste Confidence Decision Update," 73 *Federal Register* 59551, October 9, 2008.

[29] David L. Lovell, Wisconsin Legislative Council Staff, *State Statutes Limiting the Construction of Nuclear Power Plants*, October 5, 2006.

[30] *Pacific Gas & Electric Co. v. State Energy Resources Conservation and Development Commission*, 461 U.S. (190 1983).

[31] DOE, *Need for a Second Repository,* op. cit., p. 13.

[32] Department of Energy, Office of Civilian Radioactive Waste Management, *Final Supplemental Environmental Impact Statement for a Geologic Repository for the Disposal of Spent Nuclear Fuel and High-Level Radioactive Waste at Yucca Mountain, Nye County, Nevada*, DOE/EIS-0250F-S1, Washington, DC, June 2008, p. S-50, http://www.rw.doe.gov/ym_repository/seis/index.shtml.

[33] U.S. Congress, House Committee on Commerce, Subcommittee on Energy and Power, *The Nuclear Waste Policy Act of 1997*, hearing on H.R. 1270, 105th Cong., 1st sess., April 29, 1997, Serial No. 105-27 (Washington: GPO, 1997), p. 125.

[34] DOE, *Final Supplemental Environmental Impact Statement,* op. cit., p. S-51.

[35] International Nuclear Recycling Alliance, *Integrated U.S. Used Fuel Strategy*, May 1, 2008, p. 10.

[36] EnergySolutions, *GNEP Deployment Studies: Overall Summary Report*, Richland, WA, May 19, 2008, p. 2-2.

[37] Allison M. Macfarlane, Rodney C. Ewing, et al., *Uncertainty Underground: Yucca Mountain and the Nation's High- Level Nuclear Waste* (Cambridge, MA: MIT Press, 2006), p. 407.

[38] John Fleck, "WIPP Oversight Bureau Planned," *Albuquerque Journal*, June 19, 2004.

[39] http://www.nmenv.state.nm.us/doe_oversight

[40] National Academy of Sciences Board on Radioactive Waste Management, *Safety and Security of Commercial Spent Nuclear Fuel Storage: Public Report*, Washington, DC, 2005, p. 6.

[41] DOE, *Report to Congress on Interim Storage*, op. cit., p. 14.

[42] Department of Energy, Office of Civilian Radioactive Waste Management, *Monitored Retrievable Storage Submission to Congress*, DOE/RW-0035, Washington, DC, February 1986.

[43] DOE, *Report to Congress on Interim Storage*, op. cit., p. 14.

[44] Department of Energy, "DOE Holds License for Colorado Spent Fuel Facility," press release, June 28, 1999, http://newsdesk.inl.gov/press_releases/1999/DOE_Holds_Licen.htm.

[45] DOE, *Report to Congress on Interim Storage*, op. cit., p. 6.

[46] 42 U.S.C. § 2011 et seq., 10 CFR Part 72.

[47] Nuclear Regulatory Commission, *License for Independent Storage of Spent Nuclear Fuel and High-Level Radioactive Waste SNM-2513*, February 21, 2006.

[48] Private Fuel Storage, LLC, *Frequently Asked Questions: Financial Accountability*, http://www.privatefuelstorage.com/faqs/faqs.html.

[49] Bureau of Indian Affairs, *Record of Decision for the Construction and Operation of an Independent Spent Fuel Storage Installation (ISFSI) on the Reservation of the Skull Valley Band of Goshute Indians (Band) in Tooele County, Utah*, September 7, 2006.

[50] Bureau of Land Management, *Record of Decision Addressing Right-of-Way Applications U 76985 and U 76986 to Transport Spent Nuclear Fuel to the Reservation of the Skull Valley Band of Goshute Indians*, September 7, 2006.

[51] *Skull Valley Band of Goshute Indians and Private Fuel Storage, LLC, v. James E. Cason et al.* (U.S. District Court for the District of Utah, Central Division 2007).

[52] Todd D. Lovinger, "Utah Challenges License Issuance to PFS," *LLW Forum News Flash*, March 12, 2006.

[53] National Academy of Sciences, National Research Council, *Nuclear Wastes: Technologies for Separations and Transmutation*, Washington, DC, August 1995, p. 17.

[54] Peter R. Orszag, *Costs of Reprocessing Versus Directly Disposing of Spent Nuclear Fuel*, Congressional Budget Office, Statement Before the Senate Committee on Energy and Natural Resources, Washington, DC, November 17, 2007, http://www.cbo.gov/ftpdocs/88xx/ doc8808/ 11-14-NuclearFuel.pdf.

[55] Department of Energy, Office of Nuclear Energy, *Draft Global Nuclear Energy Partnership Programmatic Environmental Impact Statement*, DOE/EIS-0396, Washington, DC, October 2008, http://www.gnep.energy.gov/ peis.html.

[56] Electric Power Research Institute, *Projected Waste Packages Resulting From Alternative Spent-Fuel Separation Processes*, EPRI NP-7262 Project 3030 Final Report, Palo Alto, CA, April 1991, pp. 5-3.

[57] International Nuclear Recycling Alliance, *Presentation for GNEP Deployment Studies*, April 2008, p. 6, http://www.gnep.energy

[58] David Kramer, "Report by SRS Contractor Appears to Advocate Reprocessing at Site," *Inside Energy/with Federal Lands*, January 8, 1996, p. 8.

[59] International Nuclear Recycling Alliance, *op. cit.*, p. 43.

[60] Steven Nadis, "The Sub-Seabed Solution," *The Atlantic Monthly Digital Edition*, October 1996.

[61] Office of Technology Assessment, *Staff Paper on the Subseabed Disposal of High-Level Radioactive Waste*, Washington, DC, May 1986, p. 3.

[62] Organization for Economic Cooperation and Development, Nuclear Energy Agency, *Feasibility of Disposal of High- Level Radioactive Waste into the Seabed, Overview of Research and Conclusions*, Volume 1, Paris, 1988, p. 60.

[63] *1996 Protocol to Convention on Prevention of Marine Pollution by Dumping of Wastes*, Treaty Doc. 110-5, September 4, 2007.

[64] Atomic Energy Commission, *Draft Environmental Statement on Management of Commercial High Level and Transuranium-Contaminated Radioactive Waste*, WASH-1539, September 1974, pp. 5.3-6.

[65] Jonathan Coopersmith, "Nuclear Waste in Space?," *The Space Review*, August 22, 2005, http://www.thespacereview.com/article/437/1.

[66] Atomic Energy Commission, op. cit., p. 5.3.

[67] Luther J. Carter, *Nuclear Imperatives and Public Trust: Dealing with Radioactive Waste* (Washington, DC: Resources for the Future, 1987), p. 198.

[68] Department of Energy, *Recommendation by the Secretary of Energy Regarding the Suitability of the Yucca Mountain Site*, February 2002, p. 4, http://www.ocrwm.doe.gov/ym_repository/sr/sar.pdf.

[69] Department of Energy, Office of Civilian Radioactive Waste Management, *Draft Mission Plan for the Civilian Radioactive Waste Management Program*, DOE/RW-0005 DRAFT Volume 1, Washington, DC, April 1984, pp. 3-A23.

[70] Carter, op. cit., p. 410.

[71] Mary Louise Wagner, "DOE Decision to Halt Second Repository Program Could Derail Entire Waste Act," *NuclearFuel*, June 2, 1986, p. 7.

[72] U.S. Congress, Senate Committee on Appropriations, *Energy and Water Development Appropriation Bill, 1987*, Report to accompany H.R. 5162, 99th Cong., 2nd sess., September 15, 1986, S.Rept. 99-441 (Washington: GPO, 1986), p. 157.

[73] Department of Energy, Office of Civilian Radioactive Waste Management, *A Multiattribute Utility Analysis of Sites Nominated for Characterization for the First Radioactive-Waste Repository—A Decision-Aiding Methodology*, DOE/RW-0074, Washington, DC, May 1986, Chapter 5.

[74] Sen. Quentin Burdick, "Nuclear Waste Provisions," Remarks in the Senate, *Congressional Record*, vol. 133, part 26 (December 21, 1987), p. 37697.

[75] State of Nevada's Petition to Intervene as a Full Party in the Matter of Docket No. 63-00 1 (High Level Waste Repository), before the Nuclear Regulatory Commission, December 19, 2008. http://www.state.nv.us/nucwaste/ news2008/pdf/nv08 1219nrc.pdf

[76] Elaine Hiruo, "Funding Prospects Look Bleak for Yucca Project," *NuclearFuel*, January 12, 2009, p. 11.

[77] Rep. Morris Udall, House Debate, *Congressional Record*, vol. 133, part 26 (December 21, 1987), p. 37068.

[78] Edward F. Sproat III, Director, DOE Office of Civilian Radioactive Waste Management, speech to the Center for Strategic and International Studies, November 6, 2008.

[79] Carter, op. cit., p. 177.

[80] Carter, op. cit., p. 188.

[81] Shawn Terry, "Waste Isolation Pilot Plant Opens Doors," *Inside Energy/with Federal Lands*, March 29, 1999, p. 1.

[82] Rick Michal, "James Conca: On WIPP and Other Things Nuclear," *Nuclear News*, February 2008, p. 44.

[83] D.J. Clayton, "Effects of Heat Generation on Nuclear Waste Disposal in Salt," American Geophysical Union Fall Meeting, abstract #H53A-1010, 2008, http://adsabs.harvard.edu/abs/2008AGUFM.H53A1010C.

[84] "New Mexico Bars High-Level Waste From Carlsbad Salt Caverns," *Environment News Service*, November 4, 2004.

In: U.S. Energy and the Environment… ISBN: 978-1-61668-017-6
Editors: Roland H. Terrison pp.49-80

Chapter 3

U.S. Energy: Overview and Selected Facts and Numbers

Carl E. Behrens and Carol Glover

Summary

Energy supplies and prices are major economic factors in the United States, and energy markets are volatile and unpredictable. Thus, energy policy has been a recurring issue for Congress since the first major crisis in the 1970s. As an aid in policy making, this chapter presents a current and historical view of the supply and consumption of various forms of energy.

The historical trends show petroleum as the major source of energy, rising from about 38% in 1950 to 45% in 1975, then declining to about 40% in response to the energy crisis of the 1970s. Significantly, the transportation sector has been and continues to be almost completely dependent on petroleum, mostly gasoline. The importance of this dependence on the volatile world oil market was revealed over the past five years as perceptions of impending inability of the industry to meet increasing world demand led to relentless increases in the prices of oil and gasoline. With the downturn in the world economy and a consequent decline in consumption, prices collapsed, but the dependence on imported oil continues as a potential problem.

Natural gas followed a similar pattern at a lower level, increasing its share of total energy from about 17% in 1950 to more than 30% in 1970, then declining to about 20%. Consumption of coal in 1950 was 35% of the total, almost equal to oil, but it declined to about 20% a decade later and has remained at about that proportion since then. Coal currently is used almost exclusively for electric power generation.

Nuclear power started coming online in significant amounts in the late 1960s. By 1975, in the midst of the oil crisis, it was supplying 9% of total electricity generation. However, increases in capital costs, construction delays, and public opposition to nuclear power following the Three Mile Island accident in 1979 curtailed expansion of the technology, and many construction projects were cancelled. Continuation of some construction increased the nuclear share of generation to 20% in 1990, where it remains currently. The first new reactor license applications in nearly 30 years were recently submitted, but no new plants are currently under construction or on order.

Construction of major hydroelectric projects has also essentially ceased, and hydropower's share of electricity generation has gradually declined, from 30% in 1950 to 15% in 1975 and less than 10% in 2000. However, hydropower remains highly important on a regional basis.

Renewable energy sources (except hydropower) continue to offer more potential than actual energy production, although fuel ethanol has become a significant factor in transportation fuel, and wind power has recently grown rapidly. Conservation and energy efficiency have shown significant gains over the past three decades and offer encouraging potential to relieve some of the dependence on imports that has caused economic difficulties in the past, as well as the present.

After an introductory overview of aggregate energy consumption, this chapter presents detailed analysis of trends and statistics regarding specific energy sources: oil, electricity, natural gas, coal and renewable energy. A section on trends in energy efficiency is also presented.

INTRODUCTION

Tracking changes in energy activity is complicated by variations in different energy markets. These markets, for the most part, operate independently, although events in one may influence trends in another. For instance, oil price movement can affect the price of natural gas, which then plays a significant role in the price of electricity. Since aggregate indicators of total energy production and consumption do not adequately reflect these complexities, this compendium focuses on the details of individual energy sectors. Primary among these are oil, particularly gasoline for transportation, and electricity generation and consumption. Natural gas is also an important energy source, for home heating as well as in industry and electricity generation. Coal is used almost entirely for electricity generation, nuclear and hydropower completely so.

Renewable sources (except hydropower) continue to offer more potential than actual energy production, although fuel ethanol has become a significant factor in transportation fuel, and wind power has recently grown rapidly. Conservation and energy efficiency have shown significant gains over the past three decades, and offer encouraging potential to relieve some of the dependence on imports that has caused economic difficulties in the past as well as the present.

To give a general view of energy consumption trends, **Table 1** shows consumption by economic sector—residential, commercial, transportation, and industry—from 1950 to the present. To supplement this overview, some of the trends are highlighted by graphs in **Figures 1 and 2**.

In viewing these figures, a note on units of energy may be helpful. Each source has its own unit of energy. Oil consumption, for instance, is measured in million barrels per day (mbd),[1] coal in million tons per year, natural gas in trillion cubic feet (tcf) per year. To aggregate various types of energy in a single table, a common measure, British Thermal Unit (Btu), is often used. In **Table 1**, energy consumption by sector is given in units of quadrillion Btus per year, or "quads," while per capita consumption is given in million Btus (MMBtu) per year. One quad corresponds to one tcf of natural gas, or approximately 50 million tons of coal. One million barrels per day of oil is approximately 2 quads per year. One million Btus is equivalent to approximately 293 kilowatt- hours (Kwh) of electricity. Electric power generating capacity is expressed in terms of kilowatts (Kw), megawatts (Mw, equals 1,000 Kw) or gigawatts (Gw, equals 1,000 Mw). Gas-fired plants are typically about 250 Mw, coal-fired plants usually more than 500 Mw, and large nuclear powerplants are typically about 1.2 Gw in capacity.

Table 1 shows that total U.S. energy consumption almost tripled since 1950, with the industrial sector, the heaviest energy user, growing at the slowest rate. The growth in energy consumption per capita (i.e., per person) over the same period was about 50%. As **Figure 1** illustrates, much of the growth in per capita energy consumption took place before 1970.

Table 1. U.S. Energy Consumption, 1950-2007

	Energy Consumption by Sector (Quadrillion Btu)						Consumption Per Capita (Million Btu)		
	Resid.	Comm.	Indust.	Trans.	Total	Population (millions)	Total	Resid.	Trans.
1950	6.0	3.9	16.2	8.5	34.6	152.3	227.3	39.4	55.8
1955	7.3	3.9	19.5	9.6	40.2	165.9	242.3	44.0	57.6
1960	9.1	4.6	20.8	10.6	45.1	80.7	249.6	50.2	58.7
1965	10.7	5.8	25.1	12.4	54.0	94.3	278.0	55.0	64.0
1970	13.8	8.3	29.6	16.1	67.8	205.1	330.9	67.3	78.5
1975	14.8	9.5	29.4	18.2	72.0	216.0	333.4	68.7	84.5
1980	15.8	10.6	32.1	19.7	78.1	227.2	343.8	69.5	86.7
1985	16.1	11.4	28.9	20.1	76.5	237.9	321.5	67.6	84.4
1990	17.0	13.3	31.9	22.4	84.7	249.6	339.1	68.2	89.8
1995	18.6	14.7	34.0	23.8	91.2	266.3	342.4	69.8	89.6
2000	20.5	17.2	34.8	26.6	99.0	282.2	350.7	72.6	94.1
2005	21.7	17.9	32.5	28.4	100.5	295.9	339.7	73.4	95.8
2006	20.9	17.7	32.4	28.9	99.9	298.8	334.2	69.8	96.6
2007P	21.8	18.4	32.3	29.1	101.6	301.6	336.8	72.1	96.5

Source: Energy Information Administration (EIA), Annual Energy Review 2007, Tables 2.1a and D1. Per capita data calculated by CRS.

Notes: Data for 2007 are preliminary.

Table 1 does not list the consumption of energy by the electricity sector separately because it is both a producer and a consumer of energy. For the residential, commercial, industrial, and transportation sectors, the consumption figures given are the sum of the resources (such as oil and gas) that are directly consumed plus the total energy used to produce the electricity each sector consumed—that is, both the energy value of the kilowatt-hours consumed and the energy lost in generating that electricity. As **Figure 2** demonstrates, a major trend during the period was the electrification of the residential and commercial sectors and, to a lesser extent, industry. By 2007, electricity (including the energy lost in generating it) represented about 70% of residential energy consumption, about 80% of commercial energy consumption, and about a third of industrial energy consumption.[2]

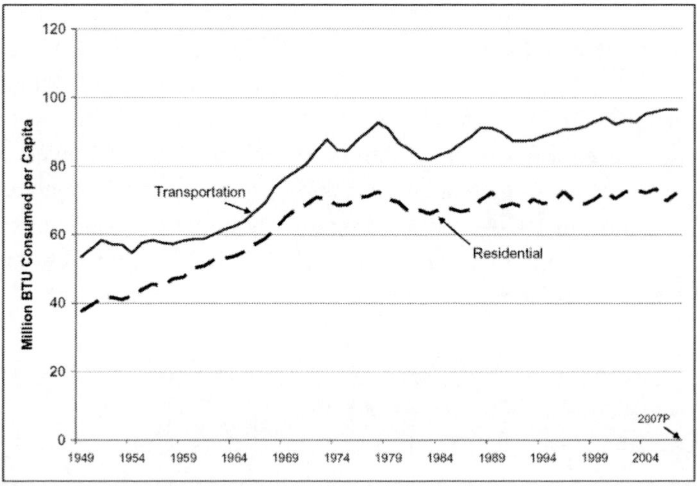

Source: Energy Information Administration (EIA), Annual Energy Review 2007, Tables 2.1a and D1. Per capita data calculated by CRS.

Notes: Data for 2007 are preliminary.

Figure 1. Per Capita Energy Consumption in Transportation and Residential Sectors, 1949-2007

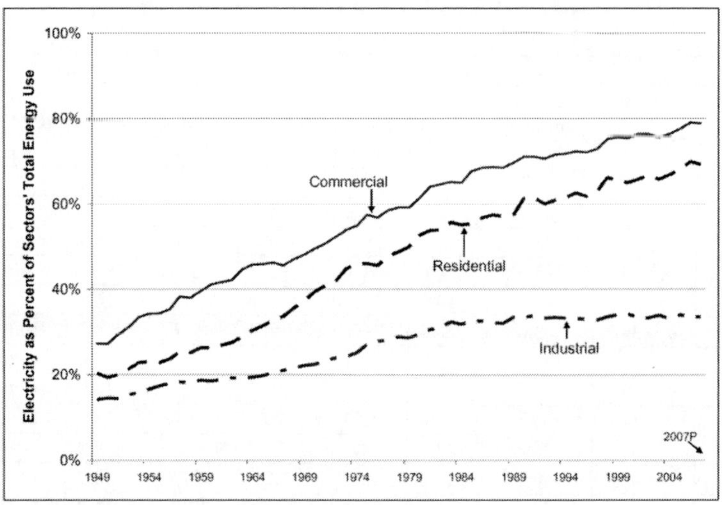

Source: Energy Information Administration (EIA), Annual Energy Review 2007, Tables 2.1a and D1. Per capita data calculated by CRS.

Notes: Data for 2007 are preliminary.

Figure 2. Electricity Intensity: Commercial, Residential, and Industrial Sectors, 1949-2007

Consumption of major energy resources—petroleum, natural gas, and coal—is presented in **Table 2** and **Figure 3**. The historical trends show that petroleum has been and continues to be the major source of energy, rising from about 38% in 1950 to 45% in 1975, then declining to about 40% in response to the energy crisis of the 1970s. Natural gas followed a similar pattern at a lower level, increasing its share of total energy from about 17% in 1950 to over 30% in 1970, then declining to about 20%. Consumption of coal in 1950 was 35% of the total, almost equal to oil, but it declined to about 20% a decade later and has remained at about that proportion since then.

OIL

About 40% of the energy consumed in the United States is supplied by petroleum, and that proportion has remained approximately the same since 1950, as the data in the previous section show. Also unchanged is the almost total dependence of the transportation sector on petroleum, mostly gasoline.

Table 2. Energy Consumption in British Thermal Units (BTU) and as a Percentage of Total, 1950-2007 (Quadrillion BTU)

	Petroleum		Natural Gas		Coal		Other		total
	Quads	% of total	Quads	% of total	Quads	% of total	Quads	% of total	
1950	13.3	38.4	6	17.3	12.3	35.5	3.0	8.6	34.6
1955	17.3	43.0	9	22.4	11.2	27.8	2.8	7.0	40.2
1960	19.9	44.1	12.4	27.5	9.8	21.7	3.0	6.5	45.1
1965	23.2	42.9	15.8	29.2	11.6	21.4	3.4	6.4	54.0
1970	29.5	43.5	21.8	32.1	12.2	18.0	4.3	6.4	67.8
1975	32.7	45.4	19.9	27.6	12.7	17.7	6.6	9.2	72.0
1980	34.2	43.8	20.4	26.1	15.4	19.7	8.3	10.6	78.1
1985	30.9	40.4	17.8	23.3	17.5	22.9	10.4	13.6	76.5
1990	33.6	39.7	19.7	23.3	19.2	22.7	12.3	14.6	84.7
1995	34.6	37.9	22.8	25.0	20.2	22.1	13.9	15.3	91.2
2000	38.4	38.8	23.9	24.1	22.7	22.9	14.2	14.4	99.0
2005	40.4	40.2	22.6	22.5	22.8	22.7	14.7	14.6	100.5
2006	40.0	40.0	22.2	22.2	22.5	22.5	15.2	15.2	99.9
2007P	39.8	39.2	23.6	23.3	22.8	22.4	15.4	15.1	101.6

Source: EIA, *Annual Energy Review 2007*, Table 1 .3.

Notes: Percentages calculated by CRS. Other includes nuclear and renewable energy. Data for 2007 are preliminary.

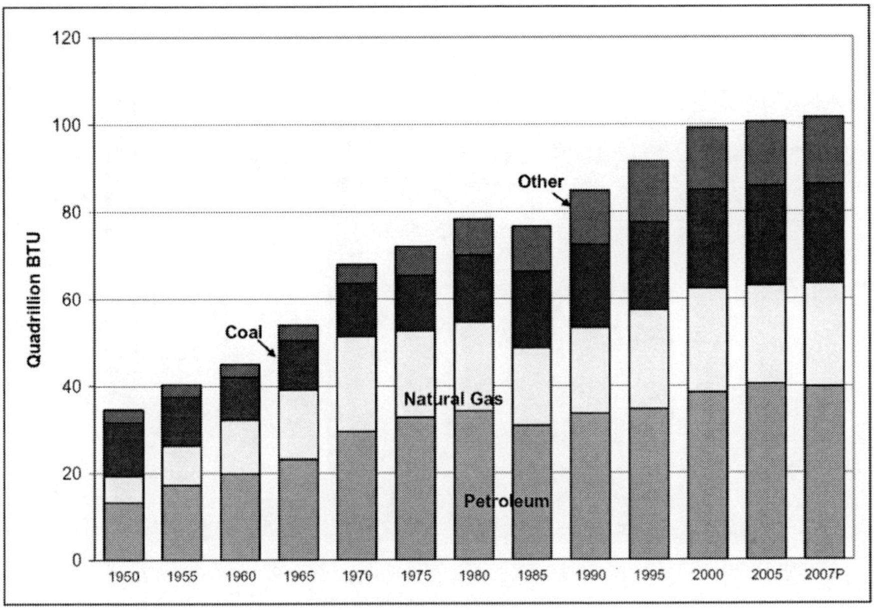

Source: EIA, *Annual Energy Review 2007*, Table 1.3.
Notes: "Other" includes nuclear and renewable energy. Data for 2007 are preliminary.

Figure 3. U.S. Energy Consumption, 1950-2007

The perception that the world is on the verge of running out of oil, widespread during the 1970s, has changed, however. The rapid price increases at that time, aided by improved exploration and production technology, stimulated a global search for oil and resulted in the discovery of large amounts of new reserves. Indeed, as concerns about tightening supply and continually increasing prices were at a peak, proven reserves actually increased by about 50% between 1973 and 1990. Some of the increase was in the Western Hemisphere, mostly in Mexico, but most was located in the region that already dominated the world oil market, the Middle East. With prices essentially steady during the 1 990s, the search for oil slowed, but additions to reserves during the decade exceeded the amount of oil pumped out of the ground. By 2003, improved technology for retrieving petroleum from oil sands in Canada and, to a lesser extent, from heavy oil in Venezuela led to significant production from these resources, and by 2005, approximately 200 billion barrels of resources from oil sands and heavy oil were added to the total of proven world reserves, 20% of the total 1991 figure. These trends are illustrated in **Figure 4**.

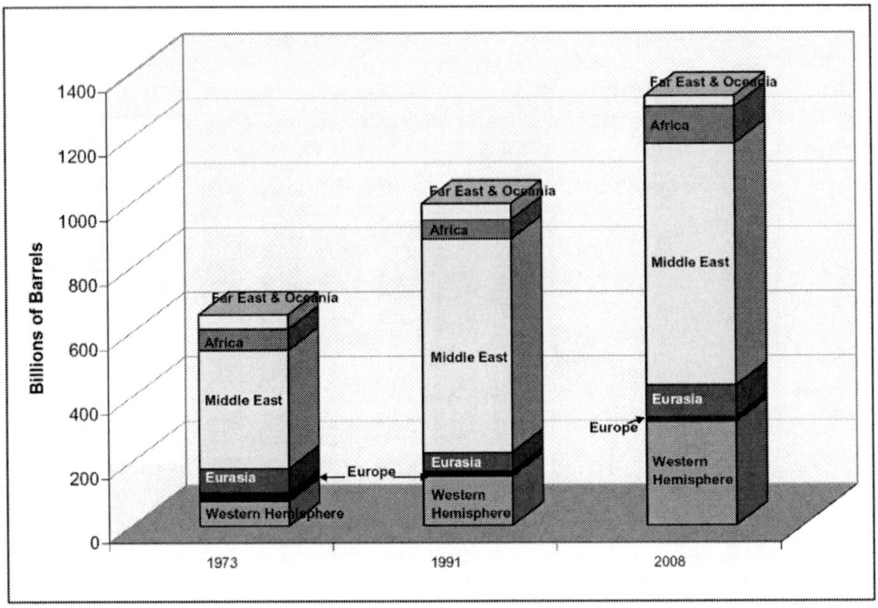

Source: EIA, International Energy Annual (IEA) 1990, Table 32 and IEA 2007 Table 8.1 Table of World Proved Oil and Natural Gas Reserves, Most Recent Estimates. (data is from Oil and Gas Journal and is not certified by EIA, except for the data for the United States in the Western Hemisphere category).

Notes: The categories "Eastern Europe and Former Soviet Union" and "Western Europe," in the data for 1973 and 1991, were changed to "Eurasia" and "Europe" respectively for 2005. Seven countries (Albania, Bulgaria, Czech Republic, Hungary, Poland, Romania, and Slovakia) were moved from the former to the latter.

Figure 4. World Crude Oil Reserves, 1973, 1991 and 2008

Petroleum Consumption, Supply, and Imports

Consumption of petroleum by sector reflects a variety of trends (see **Table 3**). In the residential and commercial sectors, petroleum consumption grew steadily from 1950 to 1970, while accounting for about 15% of total petroleum consumption. After the price surge in the 1970s, consumption in those sectors declined, falling to less than 7% of total petroleum consumption by 1995. When oil prices surged again after 2005, consumption declined further, to about 5%. Usage in the electric power sector followed a similar but more abrupt pattern. Until 1965 only about 3% of petroleum went to power generation. In the late 1960s efforts to improve air quality by reducing emissions led utilities to convert a number of coal-fired power plants to burn oil, and many new plants were designed to burn oil or natural gas. Utilities found themselves committed to increasing dependence on oil just at the time of shortages and high prices; in 1975 almost 9% of oil consumption went for power production. Consumption then fell sharply as alternate sources became available, declining to about 2%-3% of total consumption and falling even lower after 2005 as oil prices increased sharply.

Industrial consumption of petroleum, which includes such large consumers as refineries and petrochemical industries, has remained about 25% of total consumption since 1970. As other sectors' share fell, transportation, which was a little more than half of total consumption prior to 1975, climbed to two-thirds by 2000 and continued to increase its share since then.

While petroleum consumption increased throughout the period from 1950 to the present (except for a temporary decline following the price surge of the 1 970s), U.S. domestic production peaked in 1970 (see **Table 4**). The result, as shown in **Figure 5**, was greater dependence on imported petroleum, which rose from less than 20% in 1960 to about 60% in recent years.

Table 3. Petroleum Consumption by Sector, 1950-2007 (Million Barrels per Day (MBD) and Percentage of Total)

	Residential & Commercial		Industrial		Electric		Transportation		Total
	MBD	% of total	MBD	% of total	MBD	% of total	MBD	% of total	MBD
1950	1.1	16.5%	1.8	28.0%	0.2	3.2%	3.4	51.6%	6.5
1955	1.4	16.5%	2.4	28.1%	0.2	2.4%	4.5	52.4%	8.5
1960	1.7	17.5%	2.7	27.6%	0.2	2.5%	5.1	52.4%	9.8
1965	1.9	16.6%	3.2	27.2%	0.3	2.7%	6.0	52.5%	11.5
1970	2.2	14.9%	3.8	25.9%	0.9	6.3%	7.8	52.9%	14.7
1975	1.9	11.9%	4.0	24.8%	1.4	8.5%	9.0	54.9%	16.3
1980	1.5	8.9%	4.8	28.3%	1.2	6.7%	9.5	55.8%	17.1
1985	1.3	8.6%	4.1	25.9%	0.5	3.0%	9.8	62.7%	15.7
1990	1.2	7.2%	4.3	25.3%	0.6	3.3%	10.9	64.0%	17.0
1995	1.1	6.4%	4.6	26.0%	0.3	1.9%	11.7	65.9%	17.7
2000	1.3	6.5%	4.9	24.9%	0.5	2.6%	13.0	66.1%	19.7
2005	1.2	5.8%	5.1	24.5%	0.5	2.6%	14.0	67.1%	20.8
2006	1.1	5.2%	5.1	24.8%	0.3	1.4%	14.2	68.6%	20.7
2007	1.1	5.2%	5.1	24.4%	0.3	1.4%	14.3	68.9%	20.7

Source: EIA, Annual Energy Review 2007, Tables 5.1 and 5.13a-d.

Notes: Percentages calculated by CRS.

Table 4. U.S. Petroleum Production, 1950-2007 (Million Barrels per Day)

	Crude Oil			Gas Liquids	Other	Total
	48 States	Alaska	Total			
1950	5.4	—	5.4	0.5	—	5.9
1955	6.8	—	6.8	0.8	—	7.6
1960	7	—	7	0.9	0.2	8.1
1965	7.8	—	7.8	1.2	0.2	9.2
1970	9.4	0.2	9.6	1.7	0.4	11.7
1975	8.2	0.2	8.4	1.6	0.5	10.5
1980	7	1.6	8.6	1.6	0.6	10.8
1985	7.2	1.8	9	1.6	0.6	11.1
1990	5.6	1.8	7.4	1.6	0.7	9.6
1995	5.1	1.5	6.6	1.8	0.8	9.1
2000	4.9	1.0	5.8	1.9	1.0	8.7
2005	4.3	0.9	5.2	1.7	1.0	7.9
2006	4.4	0.7	5.1	1.7	1.0	7.8
2007	4.4	0.7	5.1	1.8	1.0	7.9

Source: EIA, Annual Energy Review 2007, Table 5.1.

Notes: "Other" includes processing gain.

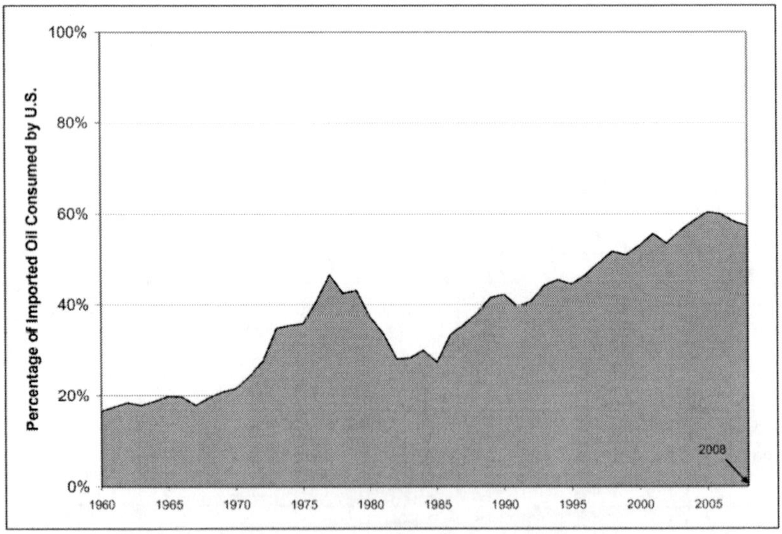

Source: EIA, Monthly Energy Review, December 2008, Table 3.3a, and Annual Energy Review 1986, Table 51.

Notes: Data for 2008 represents an 11-month average.

Figure 5. U.S. Dependence on Imported Petroleum, 1960-2007

Petroleum and Transportation

Since the transportation sector is so heavily dependent on petroleum, and uses so much of it, **Table 5** and **Figure 6** present a more detailed breakdown of the various types of petroleum used.

Aviation fuel includes both aviation gasoline and kerosene jet fuel. In 1950 aviation was almost entirely gasoline powered; by 2000 it was 99% jet fueled. The growth in flying is illustrated by the fact that aviation fuel was only 3% of petroleum consumption for transportation in 1950, but had grown to 12% in 1965 and has maintained that share since then.

Diesel fuel consumption showed a similar dramatic increase. About 6% of total petroleum consumption for transportation in 1950, it rose to 11% by 1975 and to 20% in recent years. Diesel fuel is used by a number of transportation sectors. Part of the increase involved the change of railroads from coal-fired steam to diesel and diesel-electric power. Diesel fuel is used also in the marine transportation sector, and some private automobiles are diesel-powered. The major part of diesel fuel consumption in transportation is by large commercial trucks. Total diesel fuel consumption increased from about 200,000 barrels per day in 1950 to 3.0 million barrels per day in 2007.

Table 5. Transportation Use of Petroleum, 1950-2007 (Million barrels per day)

	Aviation	Diesel Fuel	Gasoline	Other	Total
1950	0.1	0.2	2.4	0.6	3.4
1955	0.3	0.4	3.2	0.5	4.5
1960	0.5	0.4	3.7	0.4	5.1
1965	0.7	0.5	4.4	0.4	6.0
1970	1.0	0.7	5.6	0.4	7.8
1975	1.0	1.0	6.5	0.4	9.0
1980	1.1	1.3	6.4	0.7	9.5
1985	1.2	1.5	6.7	0.4	9.8
1990	1.5	1.7	7.1	0.5	10.9
1995	1.5	2.0	7.7	0.5	11.7
2000	1.7	2.4	8.4	0.5	13.0
2005	1.7	2.9	8.9	0.5	14.0
2006	1.7	3.0	9.0	0.5	14.2
2007P	1.6	3.0	9.1	0.5	14.3

Source: EIA, Annual Energy Review 2007, Table 5.13c.

Notes: Data for 2007 are preliminary.

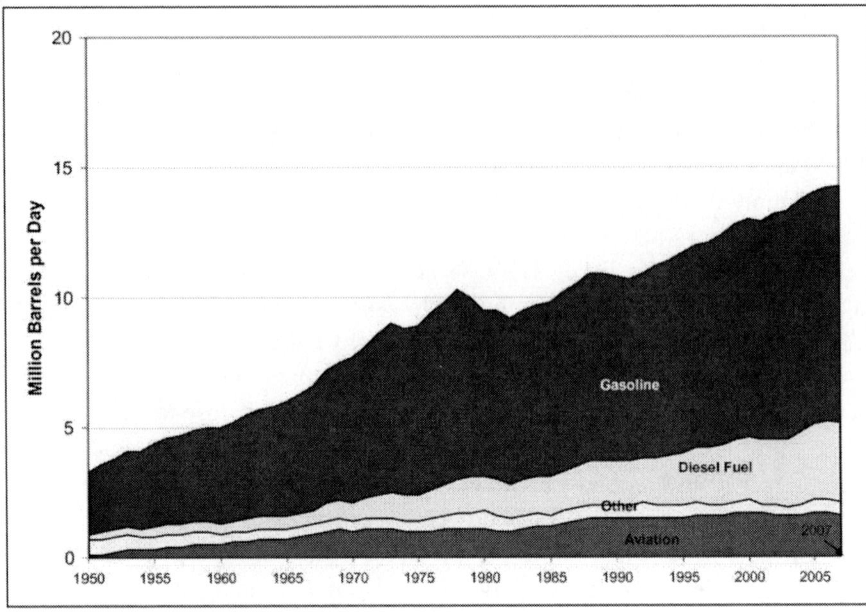

Source: EIA, Annual Energy Review 2007, Table 5.13c.
Notes: Data for 2007 are preliminary.

Figure 6. Transportation Use of Petroleum, 1950-2007

Most of the petroleum consumed in the transportation sector is motor gasoline. In 1950 it was 71% of total sector petroleum consumption, and in recent years, despite the increase in aviation fuel and diesel, it has been about 65%. Since 1950, gasoline consumption has almost quadrupled.

Of the other petroleum products consumed in the transportation sector, the largest is residual fuel oil, most of which is used in large marine transport. Consumption of residual fuel oil in the transportation sector was about 500,000 barrels in 1950, and declined gradually to about 400,000 in 2000.

Petroleum Prices: Historical Trends

Most commodity prices are typically volatile. Because oil is widely consumed, and is so important at all levels of the economy, its price is closely watched and analyzed. Especially since the 1970s, when a generally stable market dominated by a few large oil companies was broken by the Organization of Petroleum Exporting Countries (OPEC) cartel and a relatively open world market came into being, the price of crude oil has been particularly volatile. **Figure 7** and **Figure 8** show the long-term trends of crude oil and gasoline prices, in both current dollars and deflated dollars. The data for these charts do not extend to the collapse in oil prices that began in October 2008. (See "Petroleum Prices: The 2004-2008 Bubble" below.)

At the consumer level, prices of products such as motor gasoline and heating oil have reacted to price and supply disruptions in ways that have been modulated by various government and industry policies and international events. A significant and not often noted fact is that, like many commodities, the long-term trend in gasoline prices, adjusted for inflation and excluding temporary surges, has been down. As shown in **Figure 8**, the real price of gasoline peaked in 1980, then fell precipitously in the mid-1980s. The recent surge in prices brought the price above the peak of 1980 (in real dollars).

Figure 9 illustrates the proportion of the gross domestic product (GDP) dedicated to consumer spending on oil. The price surges in the 1970s pushed this ratio from about 4.5% before the Arab oil embargo to about 8.5% following the crisis in Iran late in the decade. Following that, it declined to less than 4%; during the recent run-up of prices the trend started back up again.

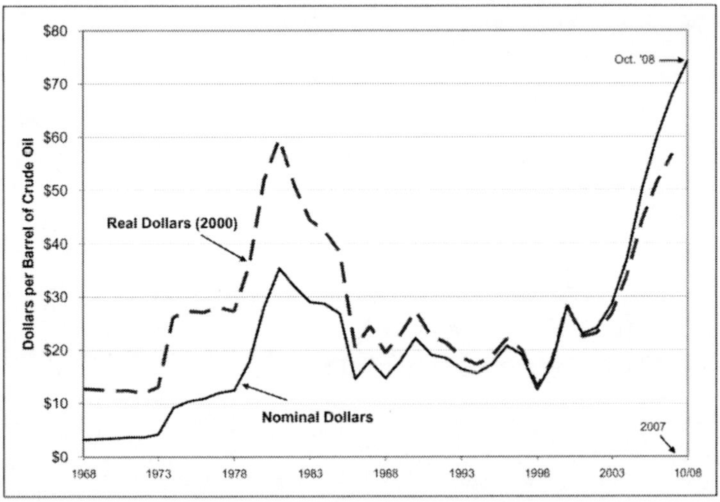

Source: EIA, Annual Energy Review 2007, Table 5.21 and Monthly Energy Review, January 2009, Table 9.1

Notes: Composite crude oil refiner acquisition cost as reported by EIA.

Figure 7. Nominal and Real Cost of Crude Oil to Refiners, 1968-2007 and October 2008

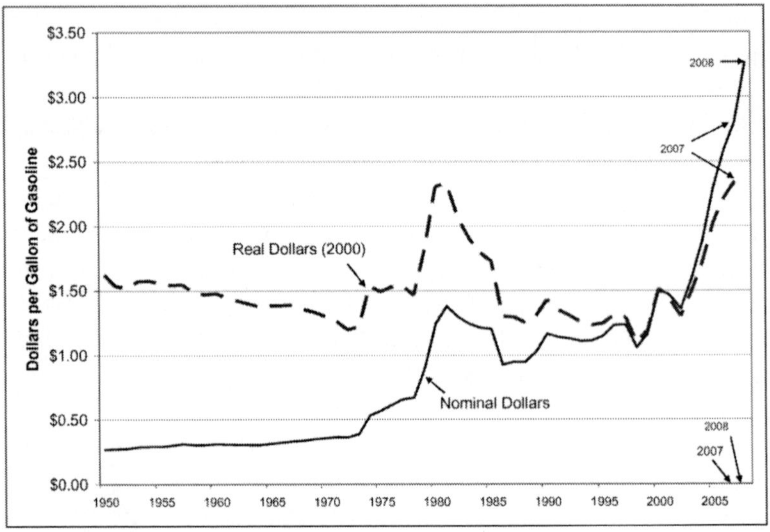

Source: EIA, Annual Energy Review 2007, Table 5.24 and Monthly Energy Review, January 2009, Table 9.4.

Notes: Average national retail price per gallon of unleaded regular gasoline, including taxes.

Figure 8. Nominal and Real Price of Gasoline, 1950-2008

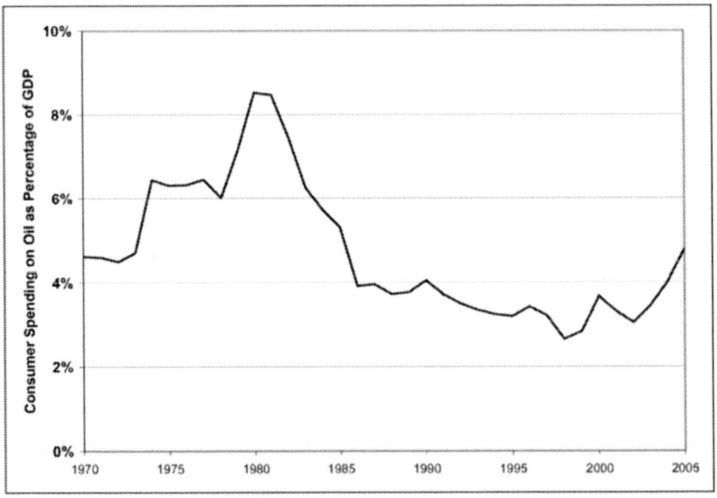

Source: EIA, Annual Energy Review 2007, Table 3.5 and Table D1 for GDP in billions of nominal dollars. Percentages calculated by CRS.

Figure 9. Consumer Spending on Oil as a Percentage of GDP, 1970-2005

Petroleum Prices: The 2004-2008 Bubble

Beginning in 2004 the world price of crude oil, and with it the price of gasoline, began to increase. Unlike the previous increases in the 1970s, there was no interruption or shortage in the supply of either petroleum or its products, except for a few months in the fall of 2005 when Hurricane Katrina shut down a major portion of U.S. refinery capacity. Nevertheless, an unexpected surge in demand for oil imports to China, added to continuing increases in demand from Europe and the United States as economies continued to grow, tightened the production capacity of the major oil producing nations and signaled that demand in the near future might not be met. In addition, turmoil in the Middle East and elsewhere, as well as the possibility of further natural disasters like Katrina, threatened supply interruptions and put further upward pressure on prices. (See **Figures 10** and **11**.)

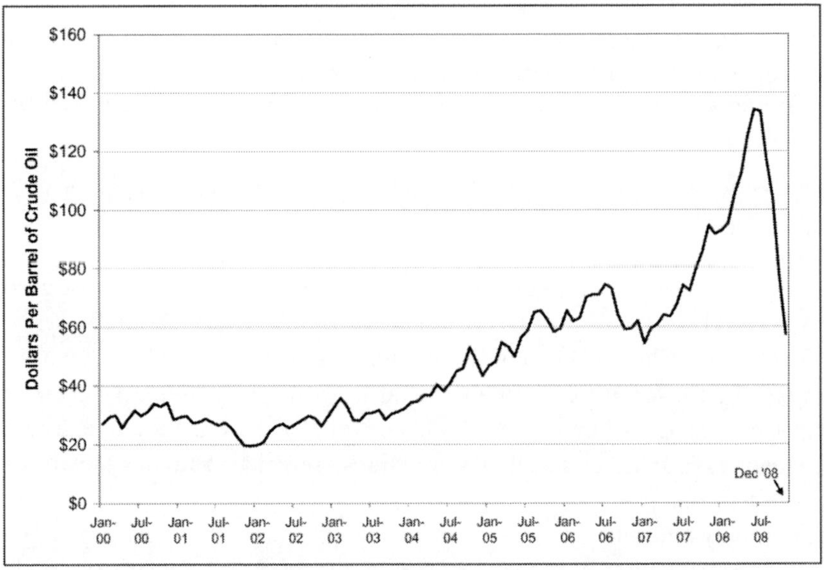

Source: EIA, NYMEX Futures Prices Crude Oil (Light-Sweet, Cushing, Oklahoma) Cushing, OK Crude Oil Future Contract 1

Notes: The futures prices shown are the official daily closing prices at 2:30 p.m. from the trading floor of the New York Mercantile Exchange (NYMEX) for a specific delivery month for each product listed.

Figure 10. Crude Oil Futures Prices.

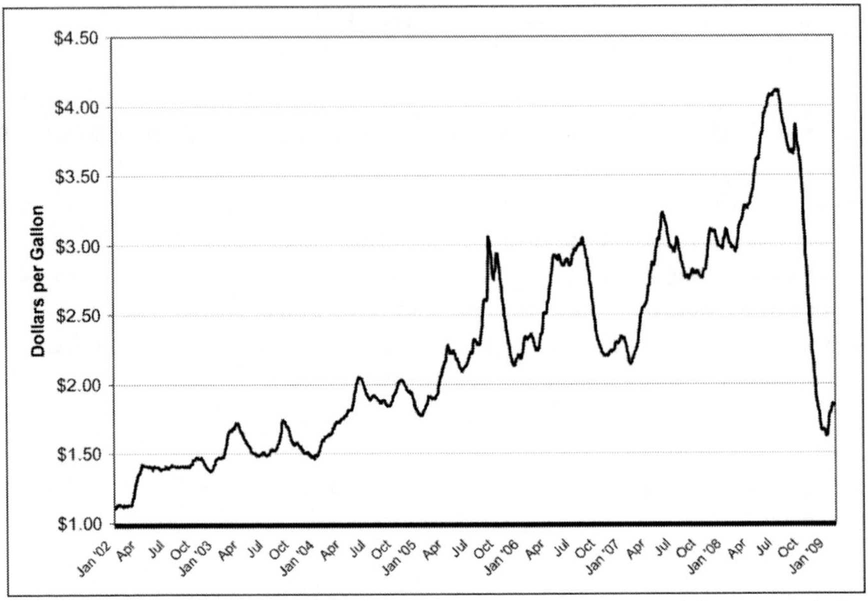

Source: Daily Fuel Gauge Report, American Automobile Association, http://www.fuelgaugereport.com, compiled by CRS.

Notes: Prices include federal, state, and local taxes. Last date above is January 29, 2009; $1.84.

Figure 11. Average Daily Nationwide Price of Unleaded Gasoline, January 2002 – January 2009

As prices continued to climb, it became apparent that demand for gasoline was relatively insensitive to its cost to the consumer. Throughout the period, as illustrated in **Figure 12**, consumption of gasoline varied seasonally but continued an upward trend on an annual basis. In the summer of 2008 crude oil prices soared far beyond the actual cost of production, and the market took on features of a classical commodities bubble, with expectations of indefinitely rising prices and participation in the market by many who would not normally enter it.

The bubble burst in October 2008 with the onset of a financial crisis in the housing and banking sectors and the evidence that consumption of gasoline was finally faltering. As the economic crisis became more acute, crude prices fell in a few months from $135 per barrel to close to $40, where they had been at the start of the run-up five years earlier.

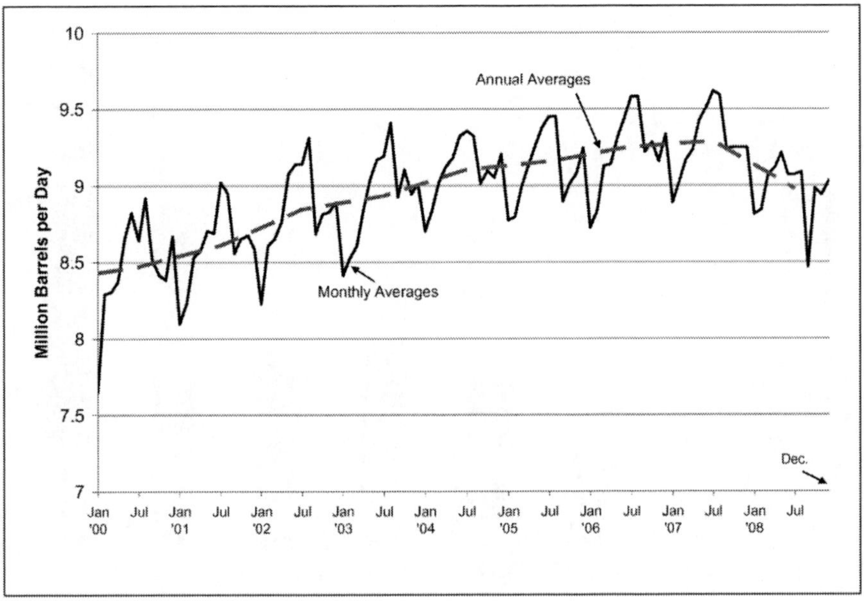

Source: EIA, Monthly Energy Review, January 2009, Table 3.5.

Figure 12. U.S. Gasoline Consumption, January 2000 – December 2008

Gasoline Taxes

The federal tax on gasoline is currently 18.4 cents per gallon. An extensive list of the gasoline and diesel fuel tax rates imposed by each state per gallon of motor fuel is maintained and updated by the American Petroleum Institute (API), "Notes to State Motor Fuel Excise and Other Tax Rates," at [http://www.api.org/policy/tax/stateexcise/upload

ELECTRICITY

While overall energy consumption in the United States increased nearly three-fold since 1950, electricity consumption increased even more rapidly. Annual power generation is ten times what it was in 1950. **Figure 13** illustrates the trend.

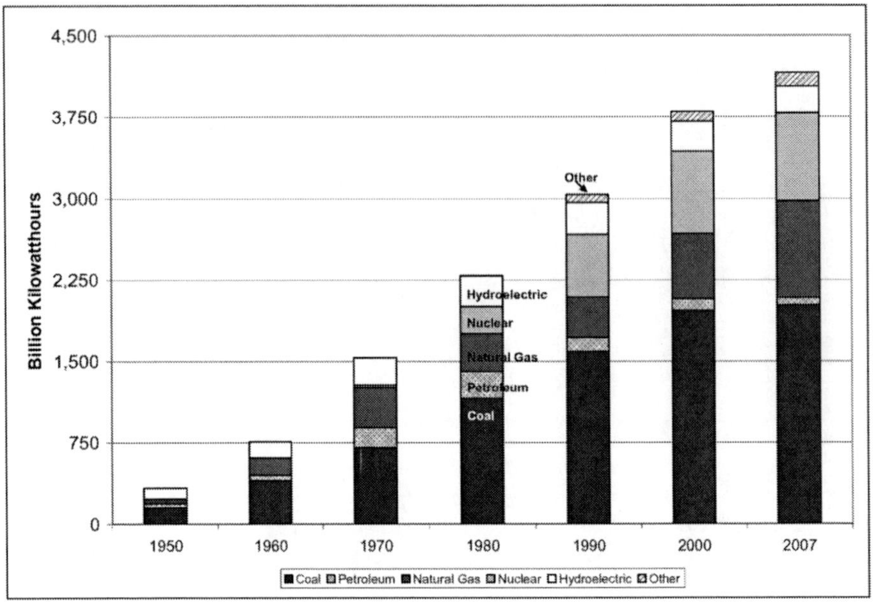

Source: EIA, Annual Energy Review 2007, Table 8.2a

Figure 13. Electricity Generation by Source, Selected Years, 1950-2007

Throughout this period, coal was used to generate about half the rapidly increasing amount of electricity consumed. Petroleum became briefly important as a source of power generation in the late 1960s because it resulted in lower emissions of air pollutants, and consumption continued in the 1970s despite the price surge because natural gas was in short supply. By the 1980s, however, oil consumption by utilities dropped sharply, and in 2007 only 1.2% of power generation was oilfired.

Natural gas generation has a more complicated history. Consumption by the electric power industry increased gradually as access by pipeline became more widespread. With the price increase in oil in the 1970s, demand for gas also increased, but interstate prices were regulated, and gas availability declined. In addition, federal energy policy viewed generation of electricity by gas to be a wasteful use of a diminishing resource. The Fuel Use Act of 1978 prohibited new power generators from using gas and set a timetable for shutting down existing gas-fired plants. Gas prices were later deregulated, resulting in increased production, and the Fuel Use Act was repealed, but in the meantime generation of electricity from gas fell from 24% in 1970 to 12% in 1985. In the 1990s gas became more popular as technology improved, and by 2000 was supplying 16% of total electric generation. Most capacity additions in the last decade have been gas-fired, as illustrated in **Figure 14**. The increased demand contributed to high prices in 2000 that were felt particularly in California.

Nuclear power started coming on line in significant amounts in the late 1960s, and by 1975, in the midst of the oil crisis, was supplying 9% of total generation. However, increases in capital costs, construction delays, and public opposition to nuclear power following the Three Mile Island accident in 1979 curtailed expansion of the technology, and many construction projects were cancelled. Continuation of some construction increased the nuclear share of generation to 20% in 1990, where it remains currently. Recently, plans have been announced for license applications for up to 30 new reactors, and several have been submitted to the Nuclear Regulatory Commission, but no new plants are currently under construction or on order.

Construction of major hydroelectric projects has also essentially ceased, and hydropower's share of electricity generation has gradually declined from 30% in 1950 to 15% in 1975 and less than 10% in 2000. However, hydropower remains highly important on a regional basis.

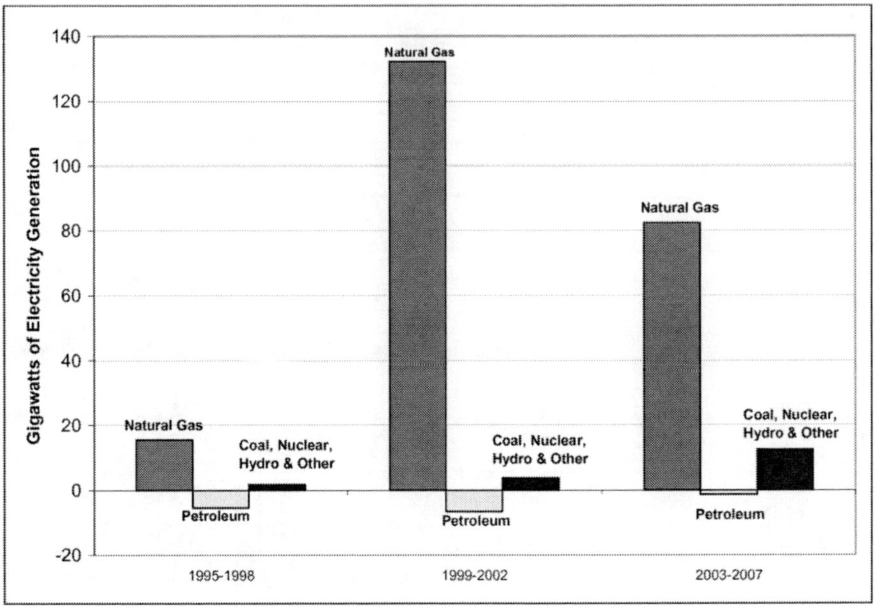

Source: EIA,Annual Energy Review 2007, Table 8.11a.

Figure 14. Changes in Generating Capacity, 1995 - 2007

Sources of power generation vary greatly by region (see **Table 6**). Hydropower in the Pacific Coast states, for instance, supplies over 40% of total generation, and natural gas almost 35%. In 2000, the combination of a drought-caused shortage of hydropower, a tightening of gas supply, and California's new electric regulatory scheme and market manipulation caused very sharp increases in electricity prices in that region. Other regions are heavily dependent on coal generation: The North Central and East South Central states, as well as the Mountain states, generate more than 60% of their electricity from coal, whereas other regions, such as New England and the Pacific Coast, use relatively little coal. The West South Central region (Arkansas, Louisiana, Oklahoma, and Texas) generates 45% of its electricity from gas. New England in the 1970s and 1980s was heavily dependent on oil-generated power; in 2005, despite an increased use of natural gas, oil produced 10% of New England's power, compared with the national average of 2.5%. By 2007, the proportion had dropped to 4.4%, and the national average to 1.2%.

Table 6. Electricity Generation by Region and Fuel, 2007

	Total Generation (billion kwh)	Percentage by					
		Coal	Petroleum	Natural Gas	Nuclear	Hydro	Other
New England	133.5	15.0%	4.4%	40.5%	27.7%	5.7%	6.8%
MiddleAtlantic	436.9	35.6%	2.2%	19.2%	34.8%	6.4%	1.8%
East North Central	670.8	69.1%	0.2%	5.3%	23.2%	0.6%	1.6%
West North Central	314.0	74.2%	0.2%	4.8%	15.4%	2.3%	3.1%
South Atlantic	836.5	53.0%	2.4%	17.0%	23.5%	1.4%	2.7%
East South Central	386.9	63.8%	0.2%	12.2%	18.7%	2.9%	2.2%
West South Central	625.7	36.8%	0.1%	45.5%	11.8%	1.2%	4.6%
Mountain	362.9	58.1%	0.1%	24.3%	7.4%	8.2%	1.9%
Pacific Contiguous	373.8	4.1%	0.1%	37.1%	11.7%	37.5%	9.5%
Pacific Noncontiguous	18.6	11.8%	55.4%	21.0%	0.0%	7.1%	4.8%
U.S. Total	9.5	48.6%	1.2%	21.5%	19.4%	6.0%	3.4%

Source: EIA, Electric Power Monthly, March 2008, Tables 1.6B, 1.7B, 1.8B, 1.10B, 1.12B, and 1.13B.

Note: "Other" includes renewables other than hydro, plus hydro from pumped storage, petroleum coke, gases other than natural gas, and other sources.

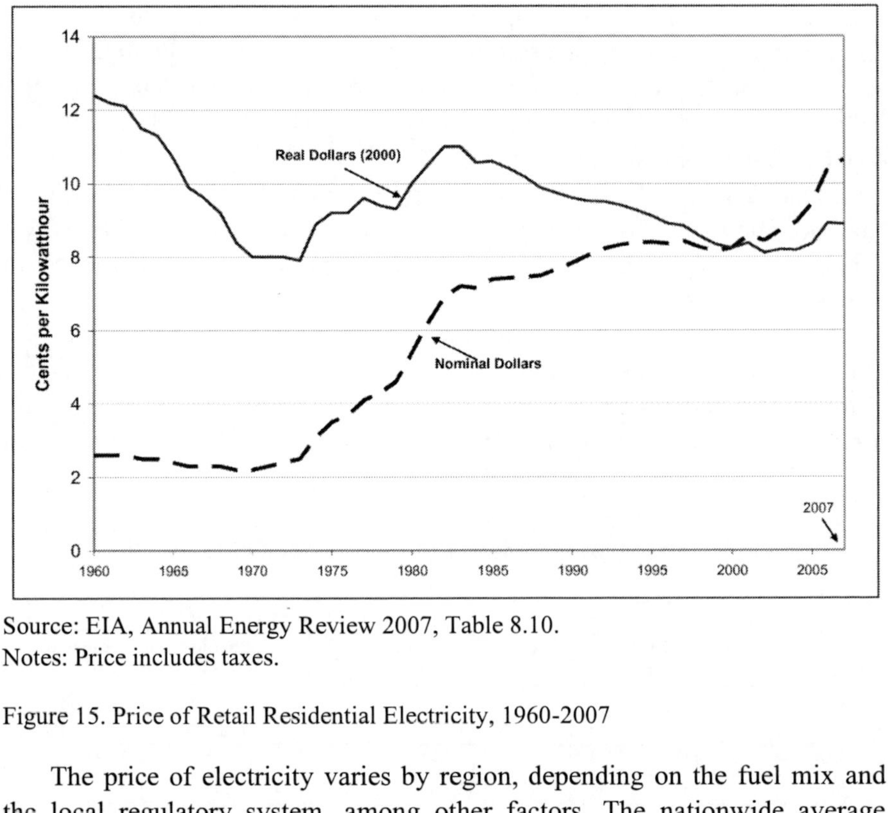

Source: EIA, Annual Energy Review 2007, Table 8.10.
Notes: Price includes taxes.

Figure 15. Price of Retail Residential Electricity, 1960-2007

The price of electricity varies by region, depending on the fuel mix and the local regulatory system, among other factors. The nationwide average retail price to residential consumers increased during the 1970s energy crises but declined starting in the 1980s, as indicated by **Figure 15**. An increase starting in 2000 resulted from the expiration in numerous regions of price caps that had been previously imposed when utilities were deregulated; the recent runup in oil and natural gas prices, and to a lesser extent in coal prices, has maintained the trend.

Other Conventional Energy Resources

Natural Gas

Consumption of natural gas was almost four times as great in 2007 as it was in 1950. Throughout the period, consumption in the residential and commercial sector grew at about the same rate as total consumption, in the range of 30% to 40% of the total. As shown in **Table 7**, consumption for electric power generation increased from about 10% in 1950 to more than 20% at the end of the century. The proportion of total gas consumption by the industrial sector declined correspondingly, from more than 50% in 1950 to about 35% in recent years.

In part because of increased demand by electric utilities, natural gas prices have become extremely volatile in recent years, as illustrated by **Figure 16**, which shows high, low, and yearly average prices for gas delivered to electricity generators.

Because rates for residential natural gas are regulated, they have been less volatile than those for electric utility consumers, although considerable seasonal fluctuations are common, as shown in **Figure 17**. The long-term trend in residential natural gas prices, both in current dollars and in constant 2006 dollars, is shown in **Figure 18**.

Table 7. Natural Gas Consumption by Sector, 1950-2007

	Total Consumption	Percent Consumed by Sector		
	trillion cubic feet (tcf)	Residential -Commercial	Industrial	Electric
1950	5.77	27.5%	59.4%	10.9%
1955	8.69	31.7%	52.2%	13.3%
1960	11.97	34.5%	48.2%	14.4%
1965	15.28	35.0%	46.5%	15.2%
1970	21.14	34.2%	43.8%	18.6%
1975	19.54	38.0%	42.8%	16.2%
1980	19.88	37.0%	41.2%	18.5%
1985	17.28	39.7%	39.7%	17.6%
1990	19.17	36.6%	43.1%	16.9%
1995	22.21	35.5%	42.3%	19.1%
2000	23.33	35.0%	39.8%	22.3%
2001	22.24	35.0%	38.1%	24.0%
2002	23.01	34.9%	37.5%	24.6%
2003	22.38	37.1%	36.9%	22.9%
2004	22.39	35.7%	37.3%	24.4%
2005	22.01	35.6%	35.0%	26.7%
2006	21.65	33.3%	35.2%	28.7%
2007P	23.06	33.5%	33.8%	29.8%

Source: EIA, Annual Energy Review 2007, Table 6.5.

Notes: Data for 2007 are preliminary. Percentages do not add to 100. The remaining amount is used by the transportation sector.

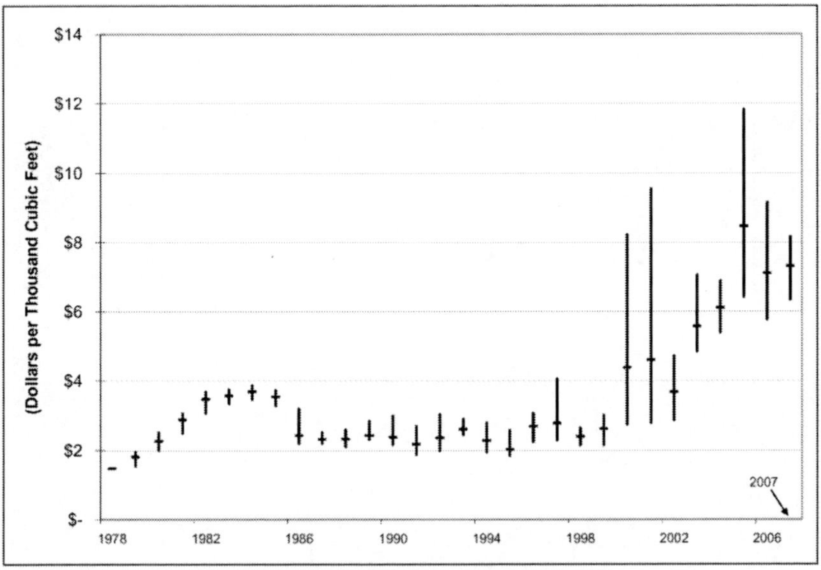

Source: EIA, Monthly Energy Review, December 2008, Table 9.11.

Figure 16. Natural Gas Prices to Electricity Generators, 1978 - 2007

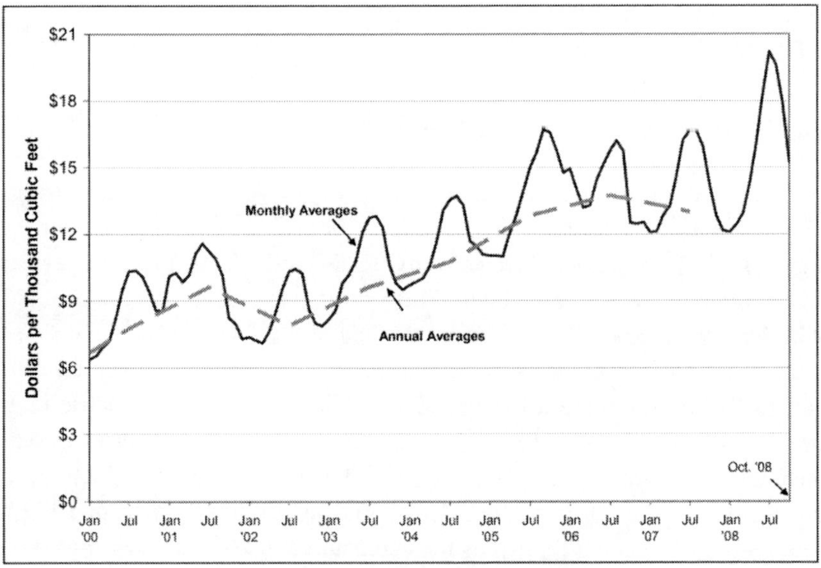

Source: EIA, Monthly Energy Review, January 2009, Table 9.11.

Figure 17. Monthly and Annual Residential Natural Gas Prices, 2000 – October 2008

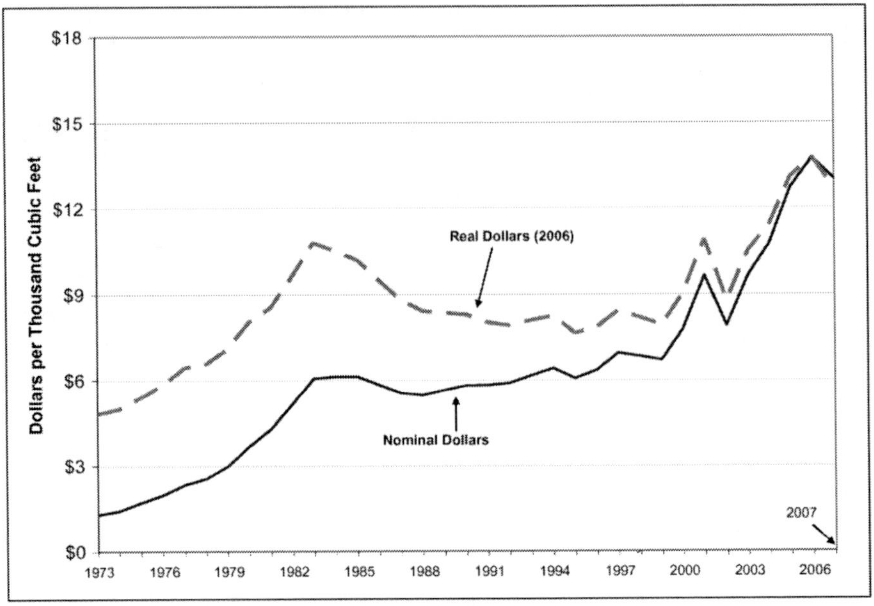

Source: EIA, Monthly Energy Review, December 2008, Table 9.11 and FY2008 Budget, Historical Tables, Table 10.1 for GDP Chained Price Index.

Figure 18. Annual Residential Natural Gas Prices, 1973-2007

COAL

Consumption of coal has more than doubled since 1950, but during that period coal as an energy source changed from a widely used resource to a single-use fuel for generating electricity. (See **Table 7**.) In 1950 the residential and commercial sector consumed almost a quarter of total coal consumed; by 1980 less than 1% of coal went to that sector. In transportation, steam locomotives (and some coal-fired marine transportation) consumed 13% of coal; by 1970 they were all replaced with diesel-burning or electric engines. Industry consumed 46% of coal in 1950; by 2000 less than 10% of coal was consumed by that sector. Meanwhile, the electric power sector, which consumed less than 20% of the half-billion tons of coal burned in 1950, used more than 90% of the billion-plus tons consumed in 2007.

Table 8. Coal Consumption by Sector, 1950-2007

	Total Consumption (million tons)	Percent Consumed by Sector			
		Residential-Commercial	Industrial	Transportation	Electric
1950	494.1	23.2%	45.5%	12.8%	18.6%
1955	447.0	15.3%	48.7%	3.8%	32.2%
1960	398.1	10.3%	44.6%	0.8%	44.4%
1965	472.0	5.4%	42.6%	0.1%	51.9%
1970	523.2	3.1%	35.7%	0.1%	61.2%
1975	562.6	1.7%	26.2%	–	72.2%
1980	702.7	0.9%	18.1%	–	81.0%
1985	818.0	1.0%	14.2%	–	84.8%
1990	904.5	0.7%	12.7%	–	86.5%
1995	962.1	0.6%	11.0%	–	88.4%
2000	1084.1	0.4%	8.7%	–	90.9%
2005	1126.0	0.4%	7.4%	–	92.1%
2006	1112.3	0.3%	7.4%	–	92.3%
2007	1128.8	0.3%	7.0%	–	92.7%

Source: EIA, Annual Energy Review 2007, Table 7.3

Notes: Data for 2007 are preliminary.

RENEWABLES

The major supply of renewable energy in the United States, not counting hydroelectric power generation, is fuel ethanol. Consumption in the United States in 2007 was 6.5 billion gallons, mainly blended into E10 gasohol (a blend of 10% ethanol and 90% gasoline). This figure represents 4.5% of the approximately 140 billion gallons of gasoline consumption in the same year. As **Figure 19** indicates, fuel ethanol production has increased rapidly in recent years, especially since the phasing out of the fuel additive methyl tertiary butyl ether (MTBE).

Another rapidly growing renewable resource is wind-generated electric power, as shown in **Figure 20**. The 300 trillion Btus of wind energy in 2006 is equivalent to approximately 88 billion kilowatt hours, about 2% of the 4,000 billion kwh of total electricity generation in that year.

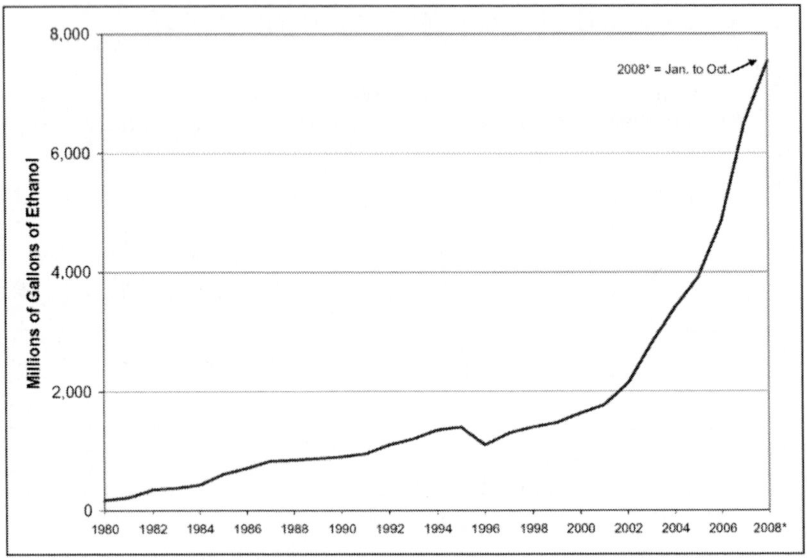

Source: Renewable Fuels Association, January 26, 2008. [http://www.ethanolrfa.org/industry/statistics/].

Notes: Data point for 2008 is for the first ten months of the year.

Figure 19. U.S. Ethanol Production, 1980-2007 and Jan.-Oct. 2008

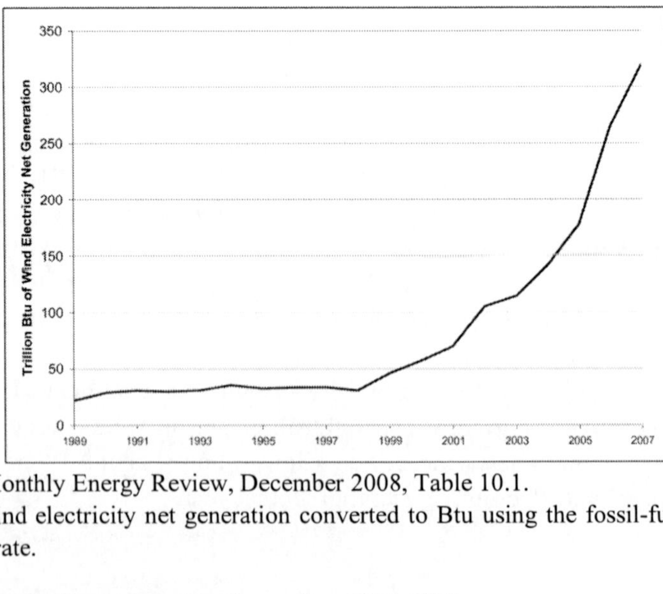

Source: Monthly Energy Review, December 2008, Table 10.1.

Notes: Wind electricity net generation converted to Btu using the fossil-fueled plants heat rate.

Figure 20. Wind Electricity Net Generation, 1989-2007

CONSERVATION AND ENERGY EFFICIENCY

Vehicle Fuel Economy

Energy efficiency has been a popular goal of policy makers in responding to the repeated energy crises of recent decades, and efforts to reduce the energy intensity of a broad spectrum of economic activities have been made both at the government and private level. Because of the transportation sector's near total dependence on vulnerable oil supplies, improving the efficiency of motor vehicles has been of particular interest. (For an analysis of legislative policies to improve vehicle fuel economy, see CRS Report R40166, *Automobile and Light Truck Fuel Economy: The CAFE Standards*, by Brent D. Yacobucci and Robert Bamberger.) **Figure 21** illustrates the trends in this effort for passenger cars and for light trucks, vans, and sport utility vehicles, as well as the general lack of improvement in heavy trucks.

Further analysis by the Environmental Protection Agency (EPA), involving the composition of the fleet as well as the per-vehicle fuel rates, indicates that light vehicle fuel economy has declined on average between 1988 and 2003. This is largely because of increased weight, higher performance, and a higher proportion of sport utility vehicles and light trucks sold. In 2003, SUVs, pickups, and vans comprised 48% of all sales, more than twice their market share in 1983. (The EPA study is available online at [http://www.epa.gov/otaq/fetrends.htm].)

Energy Consumption and GDP

A frequent point of concern in formulating energy policy is the relationship between economic growth and energy use. It seems obvious that greater economic activity would bring with it increased energy consumption, although many other factors affecting consumption make the short-term relationship highly variable. Over a longer period, for some energy-related activities, the relationship with economic growth has been essentially level. For the period from 1973 to 2003, for instance, consumption of electricity remained close to 0.45 kwh per constant dollar of GDP. Similarly, the number of miles driven by all vehicles was close to 3 miles per constant dollar of GDP throughout the same period.

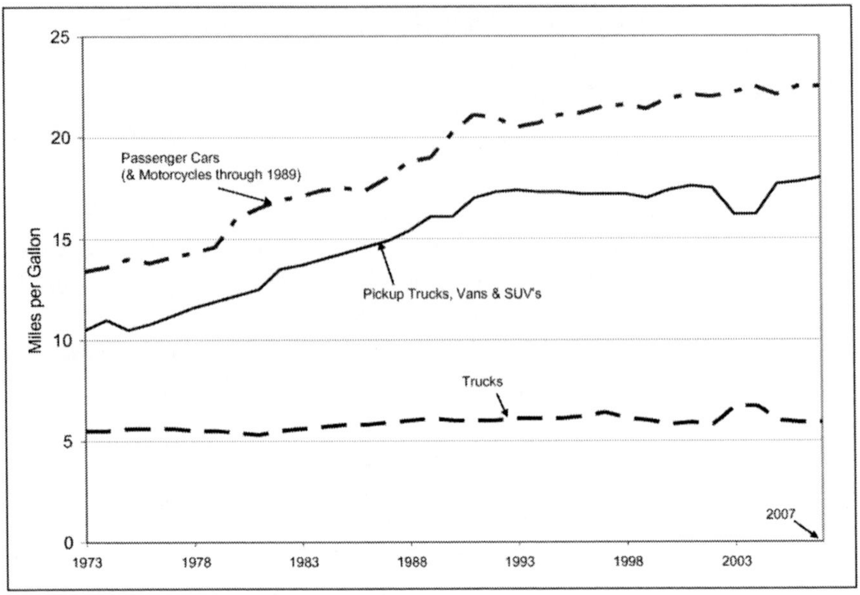

Source: EIA, Monthly Energy Review, December 2008, Table 1.8.

Figure 21. Motor Vehicle Efficiency Rates, 1973-2007

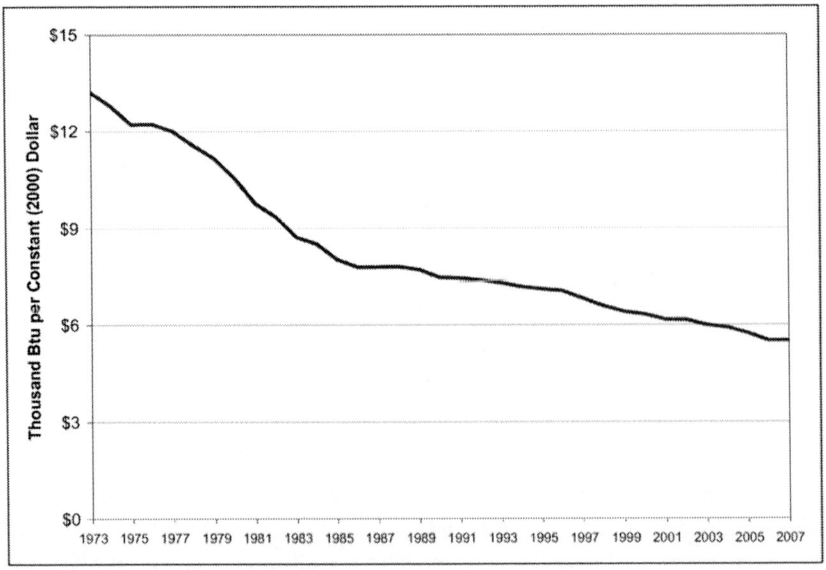

Source: EIA, Monthly Energy Review, January 2009, Table 1.7.

Figure 22. Oil and Natural Gas Consumption per Dollar of GDP, 1973 - 2007

In the case of oil and gas, however, a remarkable drop took place in the ratio of consumption to economic growth following the price spikes and supply disruptions, as illustrated in **Figure 22**. Consumption of oil and gas declined from 14,000 Btus per constant dollar of GDP in 1973 to a little more than 8,000 in 1985, and has continued to decline at a slower rate since then.

During the earlier period, oil and gas consumption actually declined 15% while GDP, despite many economic problems with inflation and slow growth, was increasing by 44% (see **Figure 22**). During the period 1987 to 2007, oil and gas consumption increased by about 25%, while GDP increased 76%.

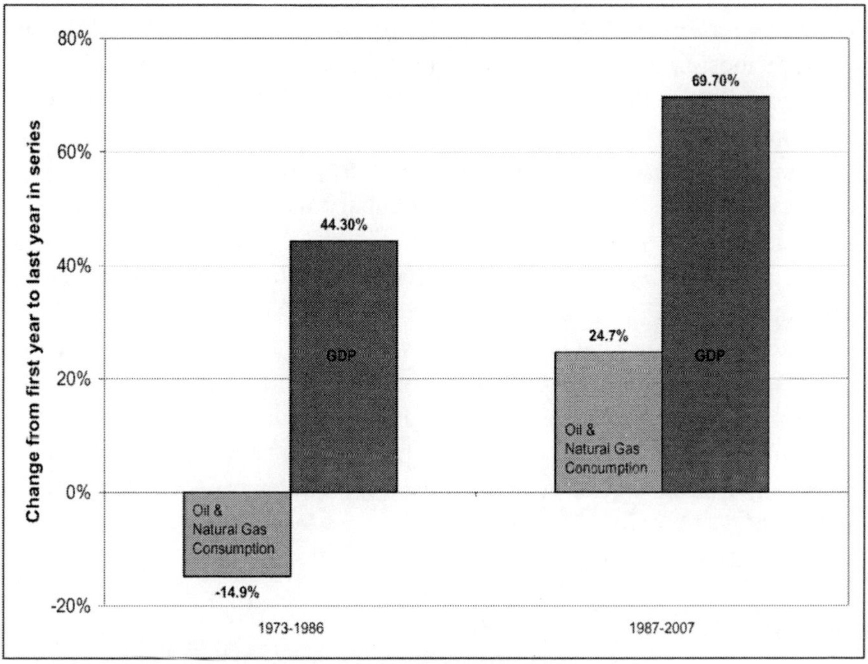

Source: EIA, Monthly Energy Review, January 2009, Table 1.8.

Notes: Percentages calculated by CRS. Percent change in oil & natural gas consumption measured in quadrillion Btu. Percent change in GDP based on billion chained (2000) dollars.

Figure 23. Change in Oil and Natural Gas Consumption and Growth in GDP, 1973-2007

Major Statistical Resources

Energy Information Administration (EIA)

EIA home page—http://www.eia.doe.gov

Most of the tables and figures in this chapter are derived from databases maintained by the Energy Information Administration (EIA), an independent agency of the Department of Energy. EIA's Website presents the complete text of its many statistical reports in PDF's and Excel files.

EIA, Publications and Reports — http://www.eia.doe.gov/bookshelf.html

EIA's most frequently requested reports include the following:

Annual Energy Review: all the historical yearly energy data across fuels

Annual Energy Outlook: energy projections out to 2030

Country Analysis Briefs: country-level energy overviews

Electric Power Monthly: monthly summary of electric power generation and capacity

International Energy Annual: international historical yearly energy data across fuels

International Energy Outlook: worldwide energy projections to 2025

Monthly Energy Review: all the latest monthly energy data across fuels

This Week in Petroleum: weekly prices & analytical summary of the petroleum industry

Weekly Petroleum Status Report: weekly petroleum prices, production & stocks data

Other Sources

Nuclear Regulatory Commission Information Digest: *http://www.nrc.gov / reading-*

Updated annually, this official NRC publication (NUREG-1350) includes general statistics on U.S. and worldwide nuclear power production, U.S. nuclear reactors, and radioactive waste

American Petroleum Institute (API): *http://api-ec.api.org/newsplashpage/ index.cfm*

The primary trade association of the oil and natural gas industry representing more than 400 members. Research, programs, and publications on public policy, technical standards, industry statistics, and regulations.

API: State Gasoline Tax Reports: *http://www.api.org/statistics*

Bloomberg.Com, Market Data: Commodities, Energy Prices: *http://www.bloomberg.com/energy*

Displays four tables:

- Petroleum ($/bbl) for crude oil. The generally accepted price for crude oil is "WTI Cushing $" which is listed third in the table.
- Petroleum (¢/gal) for heating oil and gasoline.
- Natural Gas ($/MMBtu)
- Electricity ($/megawatt hour)

This site is updated two to three times per day.

AAA's Daily Fuel Gauge Report: http://www.fuelgaugereport.com/index.asp

At-the-pump retail fuel prices for gasoline and diesel fuel. Gives average price for today, yesterday, a month ago and a year ago for wholesale and crude oil. Also displays line chart showing the averages for the previous 12 months. National, state, and metropolitan data.

International Energy Agency: http://www.iea.org

The International Energy Agency is an autonomous body within the Organization for Economic Co-operation and Development (OECD). It gathers and analyzes statistics and "disseminates information on the world energy market and seeks to promote stable international trade in energy."

A subscription is required to access most of the information on this Web site, although a limited amount of information is available to nonsubscribers. Members of Congress and their staff should contact CRS for a copy of anything that requires a subscription.

ACKNOWLEDGMENTS

Key Policy Staff

Area of Expertise	Name	Phone	E-mail
Introduction and General	Carl Behrens	7-8303	
Oil	Robert Pirog	7-6847	
Energy Taxes	Robert Pirog	7-6847	
Electricity	Stan Kaplan	7-9529	
Other Conventional Energy Sources			
Natural Gas	Robert Pirog	7-6847	
Coal	Anthony Andrews	7-6843	
Nuclear Energy	Mark Holt	7-1704	
Conservation and Energy Efficiency			
Renewable Energy	Fred Sissine	7-7039	
	Larry Parker	7-7238	
	Brent Yacobucci	7-9662	
CAFE Standards (vehicle fuel economy)	Brent Yacobucci	7-9662	
	Robert Bamberger	7-7240	
	Carol Glover	7-7353	

End Notes

[1] Further complications can result from the fact that not all sources use the same abbreviations for the various units. The Energy Information Administration (EIA), for example, abbreviates "million barrels per day" as "MMbbl/d" rather than "mbd." For a list of EIA's abbreviation forms for energy terms, see http://www.eia.doe.gov/neic/a-z/a-z_abbrev/az_abbrev.html.

[2] In calculating these percentages, "electric energy consumption" includes both the energy value of the kilowatt-hours consumed and the energy lost in generating that electricity.

In: U.S. Energy and the Environment... ISBN: 978-1-61668-017-6
Editors: Roland H. Terrison pp.81-91

Chapter 4

BIOCHAR: EXAMINATION OF AN EMERGING CONCEPT TO MITIGATE CLIMATE CHANGE

Kelsi S. Bracmort

SUMMARY

Biochar is a charcoal produced under high temperatures using crop residues, animal manure, or any type of organic waste material. Biochar looks very similar to potting soil. The combined production and use of biochar is considered a carbon-negative process, meaning that carbon is removed from the atmosphere and will not be released into the atmosphere at a later time.

Biochar has multiple potential environmental benefits, foremost the potential to sequester carbon in the soil for hundreds to thousands of years at an estimate. Studies suggest that crop yields can increase as a result of applying biochar as a fertilizer to the soil. Some contend that biochar has value as an immediate climate change mitigation strategy. Scientific experiments suggest that greenhouse gas emissions are reduced significantly with biochar application to crop fields.

Obstacles that may stall rapid adoption of biochar production systems include technology costs, system operation and maintenance, feedstock availability, and biochar handling. Biochar research and development is in its infancy. Nevertheless, interest in biochar as a multifaceted solution to agricultural and natural resource issues is growing at a rapid pace both nationally and internationally.

Past Congresses have proposed numerous climate change bills, many of which do not directly address mitigation and adaptation technologies at developmental stages like biochar. However, biochar may equip agricultural and forestry producers with numerous revenue-generating products: carbon offsets, fertilizer, and energy. A clearly defined policy medium that supports this technology has yet to emerge (e.g., soil conservation, alternative energy, climate change).

This chapter briefly describes biochar, its potential advantages and disadvantages, legislative support, and research and development activities underway in the United States and abroad.

INTRODUCTION

Biochar is a charcoal similar in appearance to potting soil. It is produced under high temperatures using crop residues, animal manure, or any type of organic waste material. Biochar is regarded by some as a product that can meet pressing environmental demands by sequestering large amounts of carbon in soil. However, little is known about how the adaptation of biochar production could successfully be implemented and what the effect would be on long-term operations in the U.S. agriculture and forestry sectors. Studies underway at federal government research institutions and in academia are focused on ensuring that biochar production systems are a practical and reliable technology for producers to adopt. The utilization of biochar is of interest to those seeking to sell or purchase carbon offsets, increase soil conservation efforts, improve crop yield, and produce alternative energy. Some contend that it will be a considerable amount of time before this technology reaches its full potential.

BIOCHAR

Biochar has the capability to both curtail greenhouse gas emissions and other environmental hazards in the near term and benefit agricultural producers as a soil amendment and source of alternative energy. Thus far, biochar use in the United States has been limited to small-scale applications reflective of the limited but growing number of researchers in this area over the last few years.

Biochar is a soil supplement that may have the potential to help mitigate global climate change through carbon sequestration in the soil. As a charcoal containing high levels of organic matter, biochar is formed from plant and crop residues or animal manure under pyrolysis conditions (**Figure 1**). Pyrolysis is the chemical breakdown of a substance under extremely high temperatures in absence of oxygen. The quantity and quality of biochar production depends on the feedstock, pyrolysis temperature, and pyrolysis processing time. A "fast" pyrolysis (~500°C) produces biochar in a matter of seconds, while a "slow" pyrolysis produces considerably more biochar but in a matter of hours.[1]

The three main outputs of a biochar production system are syngas, bio-oil, and biochar. The biochar production system is operated using energy produced by the system. Biochar production via pyrolysis is considered a carbon-negative process because the biochar sequesters carbon while simultaneously enhancing the fertility of the soil on which the feedstock used to produce the bioenergy grows (**Figure 2**).

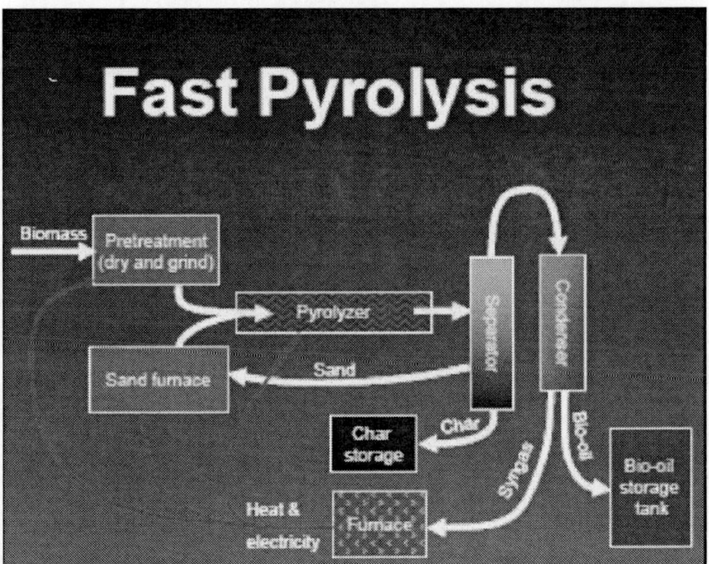

Source: U.S. Department of Agriculture (USDA) Agricultural Research Service (ARS), *ARS Biochar & Pyrolysis initiative*, 2009.

Figure 1. Biochar Production via Pyrolysis

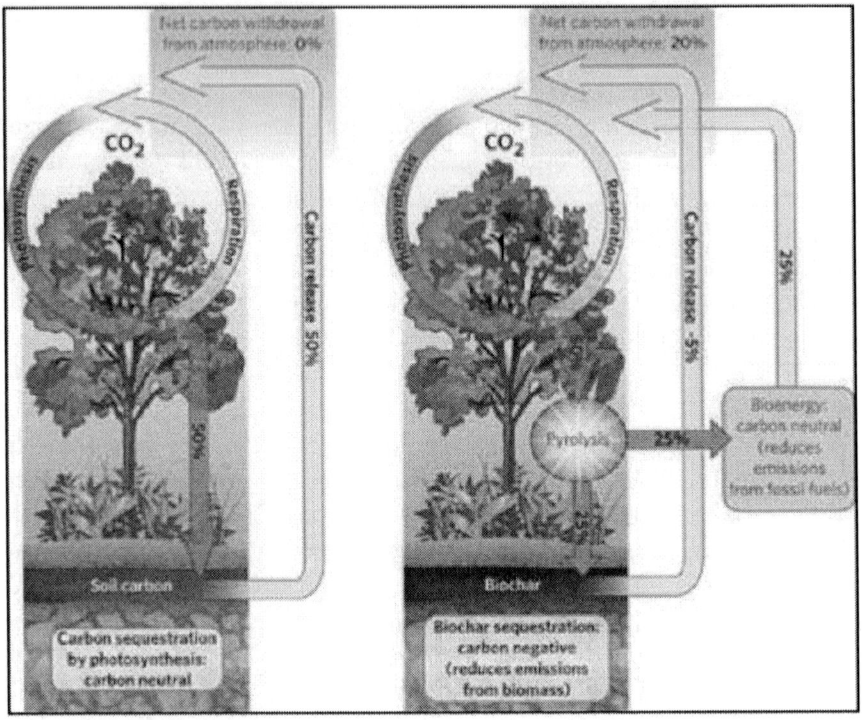

Source: J. Lehmann, "A Handful of Carbon," 2007. *Nature* 447, pp. 143-144.

Figure 2. Carbon Cycle for Soil Carbon and Biochar

Potential Advantages

Whether applied to the soil as a fertilizer or burned as an energy source (e.g., for cooking and heating), biochar provides numerous potential environmental benefits, some of which are not quantifiable. The three primary potential benefits are carbon sequestration, greenhouse gas emission reduction, and soil fertility.

Carbon sequestration

Carbon sequestration is the capture and storage of carbon to prevent it from being released to the atmosphere. Studies suggest that biochar sequesters approximately 50% of the carbon available within the biomass feedstock being pyrolyzed, depending upon the feedstock type.[2] The remaining percentage of carbon is released during pyrolysis and may be captured for energy production. Large amounts of carbon may be sequestered in the soil for long time periods (hundreds to thousands of years at an estimate),[3] but precise estimates of carbon amounts sequestered as a result of biochar application are scarce. One scientist suggests that a 250-hectare farm could sequester 1,900 tons of CO2 a year.[4]

Greenhouse gas emission reduction

Primary greenhouse gases associated with the agriculture sector are nitrous oxide (N2O) and methane (CH4). Cropland soils and grazing lands are an agricultural source of nitrous oxide emissions. Livestock manure management and enteric fermentation are leading agricultural sources of methane emissions. When applied to the soil, biochar can lower greenhouse gas emissions by substantially reducing nitrous oxide emissions.[5] Emissions of nitrous oxide, a greenhouse gas that is approximately 300 times stronger than carbon dioxide in terms of global warming potential, were reduced by 40 percent in one report.[6] Laboratory studies suggest that nitrous oxide emissions reduction from biochar-treated soil are dependent on soil moisture and soil aeration.[7] Greenhouse gas emission reductions may be 12%-84% greater if biochar is land- applied instead of combusted for energy purposes.[8]

Soil fertility

Biochar retains nutrients for plant uptake and soil fertility. The infiltration of harmful quantities of nutrients and pesticides into ground water and soil erosion runoff into surface waters can be limited with the use of biochar.[9] If used for soil fertility, biochar may have a positive impact on those in developing countries. Impoverished tropical and subtropical locales with abundant plant material feedstock, inexpensive cooking fuel needs, and agricultural soil replenishment needs could see an increase in crop yields.[10]

Potential Disadvantages

Recognizing that biochar technology is in its early stages of development, there are many concerns about the applicability of the technology in the United States. Three issues paramount to technology adoption are feedstock availability, biochar handling, and biochar system deployment. Successful implementation of biochar technology is rooted in the ability of the agricultural community to afford and operate a system that is complementary to current farming practices.

Feedstock availability

The availability of a plentiful feed supply for biochar production is an area for further study. To date, feedstock for biochar has consisted of mostly plant and crop residues, a primary domain of the agricultural community. There may be a role for the forestry community to be involved as woody biomass is deemed a cost-effective, readily available, feasible feedstock. Little is known about the advantages of using manure as a biomass feedstock. Some researchers have stated that manure-based biochar "has advantages over typically used plant-derived material because it is a by-product of another industry and in some regions is considered a waste material with little or no value. It can therefore provide a lower cost base and alleviate sustainability concerns related to using purpose-grown biomass for the process."[11]

Biochar handling

The spreading of biochar onto soil as a fertilizer is ripe for further exploration. Specifically, the ideal time to apply biochar and ensure that it remains in place once applied and does not cause a risk to human health or degrade air quality are concerns.[12] Particulate matter, in the form of dust that is hard for the human body to filter, may be distributed in abnormal quantities if the biochar is mishandled. Additionally, there are potential public safety concerns for the handling of biochar as it is a flammable substance.

Biochar system deployment

Biochar systems are designed based on the feedstock to be decomposed and the energy needs of an operation. It would be ambitious to expect a "one size fits all" standard biochar system. According to proponents, a series of mass-produced biochar systems designed for the needs of a segment of the agriculture or forestry communities might prove to be feasible (e.g., forestry community in the southeastern region, corn grower community in the midwestern region, poultry producer community in the mid-Atlantic region). Extensive deployment of biochar systems would be dependent upon system costs, operation time, collaboration with utility providers for the sale of bio-oil, and availability of information about technology reliability.

POLICY CONTEXT

Climate Change Debate

The 110th Congress deliberated over a large number of bills to address and organize climate- change mitigation efforts, including legislation for carbon offsets.[13] A carbon offset is defined as "a measurable avoidance, reduction, or sequestration of carbon dioxide (CO_2) or other greenhouse gas (GHG) emissions."[14] In the 111th Congress, carbon offsets could play a prominent role in the climate change debate. Carbon sequestration projects are one type of carbon offset. In addition to direct carbon capture and sequestration activities, the 111th Congress may consider the role of biological (indirect) sequestration, such as projects that can be implemented by agricultural producers at the field level.[15]

Congress may examine the use of biochar as an indirect carbon sequestration technology that could be used to offset carbon emissions from major emitters. In 2006, 6% of total U.S. greenhouse gas emissions were attributed to the agricultural sector.[16] While not as large as the amounts produced by some other sectors, agricultural emissions come from a large number of decentralized sources, leading many to conclude that controlling such emissions would be difficult. On the other hand, some argue that soil carbon sequestered as a result of biochar application is easily quantifiable and transparent, which may be ideal for carbon trading requirements. Others contend that ancillary benefits could include additional revenue earned by agricultural producers through the sale of carbon credits earned from biochar application or the sale of biochar as a fertilizer. Energy costs for a producer's operation may be reduced by using the alternative energy generated from the biochar production system. Additionally, some assert that the use of biochar results in higher crop yields. This could be a criterion to consider within the larger land use debate.

Farm Bill

The 110th Congress promoted biochar development through the 2008 farm bill (P.L. 110-246), which listed it under grants for High Priority Research and Extension Areas. Noted research areas include biochar production and use, co-production with bioenergy, soil enhancements, and soil carbon sequestration.

Listing biochar development as a high-priority research area in the 2008 farm bill did not authorize a specific appropriations amount. Funding for biochar development research would be determined in future appropriation bills and by the U.S. Department of Agriculture. Farm managers facing soil fertility, residue and manure management, energy efficiency, and additional revenue generation needs may benefit from a policy that further supports biochar production and use (e.g., technology practice standard, cost-share).

Long-Term Prospects

Biochar's fate as a viable component of the long-term solution to mitigate climate change by way of carbon sequestration depends upon further development by the scientific and technology transfer communities; in particular, biochar's practical application at various locations and scales using multiple feedstocks throughout the United States is an area for additional study. A policy vehicle to communicate the status of biochar technology to decision-makers and interested communities has not been identified. Natural resources conservation policy, alternative energy policy, or climate change policy are a few examples of possible policy areas. Policy that encourages academia and other institutions to conduct in-depth research and development could quicken the pace of technology deployment.

International Recognition

A series of presentations delivered at the United Nations Climate Change Conference in December 2008 has elevated interest in biochar as an immediate action response to mitigate climate change through its carbon sequestration ability.[17] Biochar's success rate as a potential clean development mechanism (CDM) mitigation technology may provide insight on its use for U.S. carbon trading purposes.[18] A CDM, monitored by the United Nations Framework Convention on Climate Change (UNFCCC), allows developed countries to invest in and receive credit for activities that reduce greenhouse gas emissions in developing countries. A formal discussion by the UNFCCC to include biochar as a CDM mitigation technology is expected to take place in 2009.

U.S. Department of Agriculture Activities

According to a U.S. Department of Agriculture (USDA) Agricultural Research Service (ARS) official, an estimated $1.1 million was allocated in 2008 toward in-house biochar research by ARS. ARS has over a dozen projects underway to analyze the utilization of fast pyrolysis to convert biomass into biochar and bio-oil at various labs nationwide. ARS estimates that the United States could use biochar to sequester 139 Tg of carbon on an annual basis if it were to harvest and pyrolyze 1.3 billion tons of biomass.[19]

Select U.S. Manufacturers

While some consider biochar research to be in its infancy, a limited number of U.S. manufacturers are selling biochar production technology to the public. CRS was not able to obtain the level of private investment in biochar technology and promotion.

Eprida, Inc.

Located in Georgia, this technology development company sells biochar production equipment. The company advocates use of both the biochar and the bio-oil produced from its patented system. Officials at Eprida, Inc., believe their technology brings value to three markets: energy, fertilizer, and carbon credit.[20]

Carbon Char Group, LLC

Agricultural grade value-added biochar is available for purchase from this New Jersey company.[21] In 2008, the company received approximately $50,000 from the USDA New Jersey Conservation Innovation Grants program (CIG) to use biochar to enhance the soil condition in sunflower fields. The CIG is a program administered by the USDA Natural Resources Conservation Service (NRCS) to encourage the development and adoption of innovative conservation technologies that work in conjunction with agricultural production.

International Activities

The International Biochar Initiative (IBI) is a nonprofit organization consisting of individuals that support a sustainable biochar production system that would remove carbon from the atmosphere and would enhance the earth's soils. IBI currently has nine developing country projects in progress to analyze cost-effective alternatives for the introduction and adoption of biochar. In Vietnam, scientists are conducting an environmental, economic, and social assessment of introducing biochar technology at the household level. The University of Tarapacá in Chile is working on a project to promote the use of biochar to stabilize and buffer soil salinity and increase water retention. Researchers in Kenya are developing a sustainable pyrolysis cook-stove and biochar system for rural agricultural households. More information concerning these projects in Africa, Asia and South America is available at http://www.biochar-international.org.

End Notes

[1] Institute for Governance & Sustainable Development and IDSG/INECE Climate Briefing Note, "Significant climate mitigation is available form biochar," December 8, 2008.

[2] Johannes Lehmann, John Gaunt, and Marco Rondon, "Bio-char Sequestration in Terrestrial Ecosystems—A Review," *Mitigation and Adaptation Strategies for Global Change*, vol. 11 (2006), pp. 403-427.

[3] Bruno Glaser, Johannes Lehmann, and Wolfgang Zech, "Ameliorating Physical and Chemical Properties of Highly Weathered Soils in the Tropics with Charcoal—A Review," *Biology and Fertility of Soils*, vol. 35 (2002), pp. 219-230.

[4] Emma Marris, "Black Is the New Green," *Nature*, vol. 442, no. 10 (August 2006), pp. 624-626.

[5] Johannes Lehmann, "Bio-energy in the Black," *Frontiers in Ecology and the Environment*, vol. 5, no. 7 (2007), pp. 381-387.

[6] Tyler Hamilton, "The Case for Burying Charcoal," *Technology Review*, April 26, 2007.

[7] Yosuke Yanai, Koki Toyota, and Masanori Okazaki, "Effects of Charcoal Addition on N2O Emissions from Soil Resulting from Rewetting Air-Dried Soil in Short-Term Laboratory Experiments," *Japanese Society of Soil Science and Plant Nutrition*, vol. 53 (2007), pp. 181-188.

[8] Johannes Lehmann, "A Handful of Carbon," *Nature*, vol. 447 (May 10, 2007), pp. 143-144.

[9] Johannes Lehmann, "Bio-energy in the Black," *Frontiers in Ecology and the Environment*, vol. 5, no. 7 (2007), pp. 381-387.

[10] Stephan M. Haefele, "Black Soil, Green Rice," *Rice Today*, April-June 2007, p. 27.

[11] K. Y. Chan, L. Van Zwieteb, and I. Meszaros et al., "Using Poultry Litter Biochars as Soil Amendments," *Australian Journal of Soil Research*, vol. 46 (2008), pp. 437-444.

[12] David A. Laird, "The Charcoal Vision: A Win-Win-Win Scenario for Simultaneously Producing Bioenergy, Permanently Sequestering Carbon, While Improving Soil and Water Quality," *Agronomy Journal*, vol. 100, no. 1 (2008), pp. 178-181.

[13] CRS Report RL34067, *Climate Change Legislation in the 110th Congress*, by Jonathan L. Ramseur and Brent D. Yacobucci.

[14] CRS Report RL34241, *Voluntary Carbon Offsets: Overview and Assessment*, by Jonathan L. Ramseur.

[15] CRS Report RL33801, *Carbon Capture and Sequestration (CCS)*, by Peter Folger; CRS Report RL33 898, *Climate Change: The Role of the U.S. Agriculture Sector*, by Renée Johnson; CRS Report RL34560, *Forest Carbon Markets: Potential and Drawbacks*, by Ross W. Gorte and Jonathan L. Ramseur.

[16] U.S. Environmental Protection Agency, *Inventory of U.S. Greenhouse Gas Emissions and Sinks: 1990-2006*, 430-R08-005, April 2008, *http://www.epa.gov/climatechange/ emissions/ usinventoryreport.html*. Unlike the U.S. Environmental Protection Agency Greenhouse Gas Inventory Report that covers all U.S. greenhouse gas emission sources and sinks, USDA only reports for the agriculture and forestry sectors. The *U.S. Agriculture and Forestry Greenhouse Gas Inventory: 1990-2005*, Technical Bulletin No. 1921, issued by the U.S. Department of Agriculture in 2008, reports the agricultural sector as having contributed approximately 481 teragrams (Tg) of carbon to the atmosphere in 2005. With the addition of forest carbon sequestration (approximately 787 Tg), the agriculture and forestry sectors combined contributed a net *sink* of approximately 306 Tg of carbon in 2005. One teragram (TG) equals one million metric tons (MMTE).

[17] "Dangerous Sea Level Rise Imminent without Large Reductions of Black Carbon and Implementation of Other Fast- Action Mitigation Strategies," *Environmental Research Web*, December 12, 2008.

[18] CRS Report RL34150, *Climate Change and the EU Emissions Trading Scheme (ETS): Kyoto and Beyond*, by Larry Parker.

[19] E-mail from Robert Fireovid, USDA ARS National Program Leader, January 8, 2008.

[20] http://www.eprida.com, 2009.

[21] http://www.carbonchar.com, 2009.

In: U.S. Energy and the Environment… ISBN: 978-1-61668-017-6
Editors: Roland H. Terrison pp.93-115

Chapter 5

ALTERNATIVE FUELS AND ADVANCED TECHNOLOGY VEHICLES: ISSUES IN CONGRESS

Brent D. Yacobucci

SUMMARY

Alternative fuels and advanced technology vehicles are seen by proponents as integral to improving urban air quality, decreasing dependence on foreign oil, and reducing emissions of greenhouse gases. However, major barriers—especially economics—currently prevent the widespread use of these fuels and technologies. Because of these barriers, and the potential benefits, there is continued congressional interest in providing incentives and other support for their development and commercialization.

Alternative fuels and advanced technology vehicles have been addressed early in the 111th Congress, as both the House and Senate versions of the American Recovery and Reinvestment Act of 2009 (H.R. 1) contained provisions supporting their development and deployment. While some of these provisions were removed in conference, the final version still contains provisions for tax incentives, federal grants and loans, and other federal support for alternative fuels and advanced vehicles.

The 111th Congress is likely to further discuss alternative fuels and advanced technology vehicles as it addresses other key topics. These include their role in any federal policy to address climate change, and their role in federal energy policy. The 111th Congress may also play an oversight role in the development of major regulations: the Environmental Protection Agency's implementation of the renewable fuel standard enacted in 2005, and expanded in 2007; the Department of Transportation's implementation of new fuel economy standards enacted in 2007; and the Department of Agriculture's implementation of a new Farm Bill enacted in 2008.

In the 110th Congress, alternative fuels and advanced technology vehicles received a good deal of attention, especially in discussions over U.S. energy security. In his January 24, 2007, State of the Union Address, President Bush called for the increased use of renewable and alternative motor fuels to 35 billion gallons annually by 2017. U.S. consumption was roughly five billion gallons in 2006. Therefore, such an initiative would mean a seven-fold increase in the use of these fuels over 11 years. On December 19, 2007, President Bush signed the Energy Independence and Security Act of 2007 (EISA, P.L. 110-140). EISA requires an increase in renewable fuel consumption to 9.0 billion gallons in 2008 and 36 billion gallons in 2022. Further within the 36- billion-gallon requirement, by 2022 the law mandates the use of 21 billion gallons of "advanced biofuels," defined as fuel derived from renewable biomass other than corn starch, with 50% lower lifecycle greenhouse gas emissions compared to petroleum fuels. The 110th Congress also enacted the Food, Conservation, and Energy Act of 2008 (2008 Farm Bill, P.L. 110-246)—which expanded and extended incentives for biofuels—as well as the Emergency Economic Stabilization Act of 2008 (EESA, P.L. 110-343)—which modified existing fuel tax credits, and established a tax credit for the purchase of plug-in vehicles.

This chapter supersedes CRS Report RL33564, *Alternative Fuels and Advanced Technology Vehicles: Issues in Congress*, by Brent D. Yacobucci.

Introduction

High levels of oil imports and high crude oil and gasoline prices in recent years have led to increased interest in the U.S. fuel supply. Recent congressional interest has focused on alternatives to petroleum, ways to improve the efficiency of the U.S. transportation sector, and ways to improve the stability and security of the petroleum supply and refining sectors.[1] From spring 2006 to summer 2008, high global oil prices (spurred by high demand)

and refinery constraints in the domestic gasoline supply pushed U.S. gasoline pump prices to historic highs.

Historically, a problem in maintaining interest in alternative fuels and vehicles has been the volatility in oil and gasoline prices. Interest tends to rise as prices rise, and decline as prices dip. Arguably, statutory policies can counterbalance dips in public interest in periods of mixed market signals as seen recently. In fall 2006 and winter 2007, gasoline prices eased somewhat; and since summer 2008, petroleum and gasoline prices have fallen dramatically.

Along with fuel prices and supply, environmental concerns, especially poor air quality and concerns over the potential effects of climate change, have further raised interest in the development of alternatives to petroleum, as well as ways to use petroleum more efficiently.

Key components of federal policies to reduce petroleum consumption include the promotion of alternatives to petroleum fuels and the promotion of more efficient vehicles. This chapter provides an overview of current issues surrounding alternative fuels[2] and advanced technology vehicles[3]— issues discussed in further detail in other CRS reports referred to in each section.

MOST RECENT DEVELOPMENTS

On February 12, 2009, House and Senate conferees filed their report on the American Recovery and Reinvestment Act of 2009 (ARRA, H.R. 1, H.Rept. 111-16). Among other provisions, ARRA modifies tax credits for alternative fuel infrastructure and plug-in vehicles, expands funding for grants to states, localities, and other entities to replace older diesel engines with cleaner diesel engines or alternative fuel engines, and establishes grants for battery manufacturers and component suppliers to develop advanced vehicle batteries and system components.

On October 3, 2008, President Bush signed the Emergency Economic Stabilization Act (EESA, P.L. 110-343). Among other provisions, EESA modified and extended key tax credits for biofuels and other alternative fuels. In addition, EESA established a tax credit for the purchase of plug-in electric vehicles.[4] On June 18, 2008, President Bush signed the Food, Conservation, and Energy Act of 2008—the 2008 "Farm Bill" (P.L. 110-246). Among other provisions, the Farm Bill modified existing biofuels tax credits; most notably, the Farm Bill lowers (starting in 2009) the tax credit for ethanol produced from corn and other conventional feedstocks, and establishes a credit for the production of advanced biofuels produced from cellulosic matter (such as trees and perennial grasses).

On December 19, 2007, President Bush signed the Energy Independence and Security Act of 2007 (EISA, P.L. 110-140).[5] EISA requires an increase in renewable fuel consumption to 9.0 billion gallons in 2008 and 36 billion gallons in 2022. Further, within the 36-billion-gallon requirement, the new law mandates the use of 21 billion gallons of "advanced biofuels," defined as fuel derived from renewable biomass other than corn starch, with 50% lower lifecycle greenhouse gas emissions compared to petroleum fuels.

Background and Analysis

Congressional Interest

Legislative background

A combination of issues—the oil crises of the 1970s, the rise in awareness of environmental issues, concerns over energy security, increasing vehicle emissions, and high gasoline prices— have spurred interest in moving the United States away from petroleum fuels for transportation and toward alternative fuels and advanced vehicle technologies.[6]

The Energy Policy Act of 1992

The 102nd Congress passed the Energy Policy Act of 1992 (EPAct 1992, P.L. 102-486). Among other provisions, this law requires the purchase of alternative fuel vehicles by federal agencies, state governments, and alternative fuel providers.[7] Under EPAct 1992, a certain percentage— which varies by the type of fleet (i.e., federal, state, or fuel provider)—of new passenger vehicles must be capable of operating on alternative fuels, including ethanol, methanol, natural gas, or propane. EPAct 1992 established a tax credit for the purchase of electric vehicles, as well as tax deductions for the purchase of alternative fuel and hybrid vehicles.

The Energy Policy Act of 2005

There was little congressional action on energy policy through the late 1990s. In light of high fuel prices in the early 2000s, continued growth in domestic and global petroleum demand, and other energy policy concerns,

Congress began working on comprehensive energy legislation in 2001. In the 107th Congress, an energy bill stalled in conference. The 108th Congress continued the debate over energy legislation. The conference report (H.Rept. 108-375) included provisions on vehicle tax credits, amendments to vehicle purchase requirements under the Energy Policy Act of 1992, a requirement that gasoline contain ethanol or other renewable fuels, and tax credits for ethanol and biodiesel fuels. However, this bill also stalled. Many of these topics were addressed in the 109th Congress by the Energy Policy Act of 2005 (EPAct 2005, P.L. 109-58), which was signed by President Bush on August 8, 2005.

The Energy Independence and Security Act of 2005

Continued pressure on energy prices and concerns over energy security after passage of EPAct 2005 led to continued discussion of energy policy in the 110th Congress. On December 19, 2007, President Bush signed the Energy Independence and Security Act of 2007 (EISA, P.L. 110-140). Among other provisions, EISA expanded the renewable fuel mandate in EPAct 2005, and significantly tightened federal fuel economy (CAFE) standards.

The 2008 Farm Bill

Biofuels—fuels produced from renewable organic matter, especially agricultural products and wastes, are seen by proponents as a key strategy for increasing energy security, promoting environmental quality, and raising farm incomes. Therefore, recent Farm Bills, especially the 2002 and 2008 Farm Bills (P.L. 107-171 and P.L. 110-246, respectively), have included titles to promote biofuels and other farm-based energy supplies. The 2002 Farm Bill established programs to promote the development of biofuels and biorefineries; the 2008 Farm Bill expanded on these programs, and expanded existing biofuels tax credits to promote the development of cellulosic fuels—fuels produced from woody or fibrous materials such as perennial grasses, fast-growing trees, and agricultural and municipal wastes.

The American Recovery and Reinvestment Act of 2009

The American Recovery and Reinvestment Act of 2009 (H.R. 1), reported by House and Senate conferees February 12, 2009, includes several key provisions supporting alternative fuels and advanced technology vehicles. These include tax credits for the purchase of small electric vehicles, grants to states, localities, and other entities to replace older diesel engines with new, clean diesel or alternative fuel engines, and grants to battery manufacturers and part suppliers to develop batteries and system components for advanced vehicles (e.g., hybrids, plug-in electric vehicles). A provisions contained in the Senate version to expand U.S. Department of Agriculture (USDA) biorefinery grants was dropped by the Conference Committee. For a more detailed comparison of the House, Senate, and Conference versions of the bill, see **Table 1**.

Other Legislation

Other laws affecting alternative fuel and advanced technology vehicles include the Energy Policy and Conservation Act (P.L. 94-163), which established fuel economy standards for passenger cars and light trucks;[8] the 1990 Amendments to the Clean Air Act (P.L. 101-549), which require cities with significant air quality problems to promote low emission vehicles; highway authorization bills, including P.L. 109-59 and P.L. 105-178, which established and reaffirmed tax incentives for ethanol and other fuels; and numerous laws that authorize federal research and development on alternative fuels, advanced technologies, and enabling infrastructure, such as alternative fuel pumps. The Emergency Economic Stabilization Act of 2008 (P.L. 110-343) modified and extended key tax incentives for biodiesel and other alterative fuels.

Current Issues

Recent events have renewed interest in alternative fuels and advanced vehicles. For example, high pump prices for gasoline and diesel fuel through summer 2008 have raised concerns over fuel conservation and energy security, including U.S. dependency on oil imports. In light of this, there is growing interest in more efficient vehicles or vehicles that abandon the use of petroleum altogether. This is especially true as the rapid growth in the sales of light trucks—these include sport utility vehicles (SUVs), mini-vans, and pickups, which tend to have lower fuel economy than passenger cars—through the mid-2000s has lowered the overall fuel economy of the new vehicle fleet. EISA requires an increase in fuel economy from passenger cars and light trucks to 35 miles per gallon (mpg) combined in 2020 from roughly 24 mpg today.

Ongoing technological developments in hybrid vehicles, ethanol fuel, fuel cells, and hydrogen fuel have raised key policy questions. These questions include whether more generous tax incentives for hybrid and/or fuel cell vehicles should be established, the costs and environmental impacts associated with production of ethanol or hydrogen as major transportation fuels, and whether research and development funds should be focused on such potentially high-risk technologies as fuel cells or on near-term technologies, such as hybrids.

Gasoline prices have spiked and the gasoline supply system has faced disruptions several times in recent years, driven by various factors, including:

- Hurricanes along the Gulf Coast in the fall of 2005;
- High crude prices, issues with refining capacity, and concerns about ethanol supply in spring 2006;
- Historic high crude oil and gasoline prices returned to historic highs in 2007 and 2008.

These price surges and supply disruptions raised congressional interest in alternatives to petroleum. Coupled with concerns over the environmental impact of petroleum and other fossil fuels, congressional interest in alternatives remains strong, even though oil and gasoline prices declined during the second half of 2008.

Fuel Tax Incentives

There are three key tax incentives for alternative fuels: (1) a tax credit for conventional ethanol of $0.45 per gallon,[9] (2) a tax credit for biodiesel and renewable diesel of $1.00 per gallon,[10] and (3) a credit of $0.50 per gallon for the retail sale of alternative fuels other than ethanol and biodiesel (e.g., LPG). In addition, there are tax credits for small ethanol and biodiesel producers ($0.10 per gallon), and a tax credit for the production of cellulosic biofuels (up to $1.01 per gallon, depending on the fuel).[11]

There is ongoing interest in tax incentives for the production and purchase of alternative fuels. Supporters of this approach argue that the market favors conventional fuels, and that the widespread infrastructure and nearly ubiquitous use of conventional fuels in automobiles makes it difficult for alternative fuels to compete without economic incentives. The American Jobs Creation Act of 2004 (P.L. 108-357) replaced a previous excise tax exemption for ethanol- blended fuels with a tax credit. The credit was valued at $0.51 per gallon in 2008 and is expected to be reduced to $0.45 per gall for 2009.

In addition to the credit for ethanol-blended gasoline, there has been interest in promoting biodiesel fuel. P.L. 108-357, and subsequent amendments, provides a tax credit of $1.00 per gallon for the sale and use of biodiesel. P.L. 109-58 expanded the credit to include "renewable diesel," which is produced from a different process than biodiesel and results in a fuel with somewhat different chemical characteristics. In recent guidance on the tax credit, the Internal Revenue Service ruled that renewable diesel includes synthetic diesel fuel produced from vegetable oils at petroleum refineries.[12] Most biodiesel producers are small plants, and many biodiesel producers are concerned that this decision could lead to a shift away from biodiesel production to renewable diesel production at large refineries. The Emergency Economic Stabilization Act of 2008 (EESA) included the following language amending the renewable diesel tax credit: "such term does not include any fuel derived from coprocessing biomass with a feedstock which is not biomass."[13] Presumably, this provision is intended to limit the production of renewable diesel eligible for the tax credit at petroleum refineries. EESA also extended the biodiesel and renewable diesel credits through the end of 2009.

Ethanol and MTBE

Outside of tax incentives, ethanol has been of key interest in recent Congresses, especially in its role as an alternative to gasoline and to MTBE (methyl tertiary butyl ether).[14] MTBE and ethanol were used (among other purposes) to meet previous Clean Air Act requirements that reformulated gasoline (RFG), sold in the nation's worst ozone nonattainment areas, contain oxygen to improve combustion. Under the RFG program, areas with "severe" or "extreme" ozone pollution (90 counties with a combined population of 64.8 million[15]) must use reformulated gas; areas with less severe ozone pollution may opt into the program as well, and many have. In all, portions of 17 states and the District of Columbia use RFG, and about 30% of the gasoline sold in the United States is RFG, according to the Environmental Protection Agency (EPA).[16]

Before amendment by the Energy Policy Act of 2005, the Clean Air Act required that RFG contain at least 2% oxygen by weight.[17] Refiners met this requirement by adding a number of ethers or alcohols, any of which contains oxygen and other elements. Until about 2003, the most commonly used oxygenate was MTBE because it was cheaper and easier to use than competing oxygenates. In 1999, 87% of RFG contained MTBE, a number reduced to about 46% in 2004, according to EPA. MTBE has also been used since the late 1 970s in non-reformulated gasoline as an octane enhancer, at lower concentrations. As a result, gasoline with MTBE has been used throughout the United States, whether or not an area has been subject to RFG requirements.

MTBE contamination creates taste and odor problems in water at very low concentrations, and some animal studies indicate MTBE may pose a cancer risk to humans. MTBE leaks, generally from underground gasoline storage tanks, have been implicated in numerous incidents of ground water contamination. For these reasons, 25 states have taken steps to ban or limit its use, according to the Renewable Fuels Association.[18] The most significant of the bans (in California and New York) took effect at the end of 2003, leading many to suggest that Congress revisit the issue to modify the oxygenate requirement and set more uniform national requirements regarding MTBE and its potential replacements, principally ethanol.

Support for eliminating the oxygenate requirement on a nationwide basis was widespread among states, the petroleum industry, and some environmental groups. In general, these stakeholders concluded that gasoline can meet the same low-emission performance standards as RFG without the use of oxygenates. But agricultural interests presented a potential obstacle to enacting legislation to remove the oxygen requirement. According to USDA, roughly 20% of the nation's corn crop was used in 2006/2007 to produce the competing oxygenate, ethanol.[19] If MTBE use were reduced or phased out, but the oxygen requirement remained in effect, ethanol use would have soared, increasing demand for corn. Conversely, if the oxygen requirement were repealed, not only would MTBE use decline, but so, likely, would demand for ethanol. Thus, some Members of Congress and governors from corn-growing states took a keen interest in MTBE legislation and related oxygenate requirements.

To help promote the market for ethanol if the oxygen standard were eliminated, a renewable fuel standard (RFS) was suggested. This would require that all gasoline contain ethanol or other renewable fuel. This concept was supported by agricultural interests, the oil industry, and some environmental groups. Opponents included states that do not produce ethanol, due to fears that the mandate could raise gasoline prices.

The Energy Policy Act of 2005 (P.L. 109-58) contains numerous MTBE and ethanol provisions. It repealed the Clean Air Act requirement to use MTBE or other oxygenates. In place of this requirement, the law established a renewable fuel standard. Under the RFS, annual gasoline supply was required to contain 7.5 billion gallons of ethanol or other renewable fuel by 2012. To prevent "backsliding" on air quality, the law requires that reductions in emissions of toxic substances achieved by RFG be maintained, and it authorizes funds for MTBE cleanup.[20] The Energy Independence and Security Act of 2007 (P.L. 110-140) expanded this mandate, requiring the use of 9.0 billion gallons of renewable fuels in transportation fuel[21] in 2008, and 36 billion gallons in 2022.

Cellulosic Biofuels[22]

Ethanol, the most significant biofuel in the United States, is usually produced from corn. However, corn is a key animal feed, and is also used for human consumption. Further, corn is a resource-intensive crop, requiring significant use of chemical fertilizers and generally grown on prime farmland. There is growing interest in developing biofuels that require less energy to produce and have a smaller environmental footprint.

Biofuels produced from cellulosic materials such as fast-growing trees, prairie grasses, or agricultural wastes are seen as a potential strategy for reducing the environmental impact of biofuels while expanding the United States' ability to displace petroleum fuels. The potential supply of these feedstocks is abundant, which is why it is expected that future expansion of the U.S. biofuels industry will be in this area.

However, breaking down cellulose and converting it into fuel requires complex chemical processing. Starches (such as corn) and sugars (such as cane sugar) are easily fermented into alcohol, while cellulose must be broken down into sugars or starches through enzymatic or thermochemical processes before fermentation. Alternatively, biomass can be converted into synthesis gas,[23] which can then be used to produce fuels. Regardless of the pathway, processing cellulose into fuels is currently prohibitively expensive relative to other conventional and alternative fuel options. Therefore, R&D has focused on lowering the costs of enzymatic and other processing techniques.

Further, questions remain about the feasibility of these fuels, as well as the ultimate environmental footprint—many of the proposed feedstocks have never been grown on a large scale. Therefore, R&D is also focused on increasing the yield of potential biofuel crops, developing harvesting techniques, and finding ways to limit the environmental impact of dedicated energy crops.

The Energy Policy Act of 2005 included provisions to promote the development of cellulosic biofuels. These include an authorization for increased research and development funding at the Department of Energy; grants, loans, and loan guarantees for the development of cellulosic biofuels;

per-gallon incentives for the first 1 billion gallons of domestic production;[24] and a mandate that gasoline contain at least 250 million gallons of cellulosic ethanol annually starting in 2013.

On December 20, 2006, President Bush signed the Tax Relief and Health Care Act of 2006 (P.L. 109-432). Among other provisions, this tax law established a 50% depreciation allowance for cellulosic ethanol plants placed in service before January 1, 2013, subject to certain limitations.

The Energy Independence and Security Act of 2007 expanded the renewable fuel mandate in EPAct 2005, and established specific requirements for "advanced biofuels"—defined as fuels produced from feedstocks other than corn starch, and with 50% lower lifecycle greenhouse gas emissions than petroleum fuels. Of the 36 billion gallons of renewable fuel required in 2022, 21 billion gallons must be advanced biofuels; within that mandate, there are specific carve-outs for cellulosic biofuels and biomass-based diesel fuels.

Ethanol Imports

Corn growers and ethanol producers are supportive of the renewable fuel standard because of its implications for higher corn and ethanol prices. However, concern over ethanol imports is growing among some stakeholders. Because of lower production costs and the availability of government incentives, ethanol prices in Brazil and some other countries can be significantly lower than in the United States. To offset the U.S. tax incentive that all ethanol (imported or domestic) receives, most imports are subject to a relatively small 2.5% ad valorem tariff, but more significantly an added duty of $0.54 per gallon. This added duty effectively negates the tax incentive for covered imports and has been a significant barrier to fuel ethanol imports.
However, under certain conditions imports of ethanol from Caribbean Basin Initiative (CBI) countries are granted duty-free status.[25] This is true even if the ethanol was produced in a non- CBI country. In this scenario, the ethanol is produced in another country (historically Brazil or a European country), dehydrated in a CBI country, then shipped to the United States. In recent years, these imports have reached as high as 5% of the U.S. ethanol market. This avenue for imported ethanol to avoid the tariff has been criticized by some stakeholders, including some Members of Congress. With the establishment of a renewable fuel standard, as well as high U.S. gasoline and ethanol prices, there may be more interest in importing ethanol, either through CBI countries or directly from ethanol producers.

In addition to the concerns over imports of duty-free ethanol from CBI countries, there is growing concern that a large portion of ethanol otherwise subject to the duties is being imported duty-free through a "manufacturing drawback."[26] If a manufacturer imports an intermediate product, then exports the finished product or a similar product, then that manufacturer may be eligible for a refund (drawback) of up to 99% of the duties paid. There are special provisions for the production of petroleum derivatives.[27] In the case of fuel ethanol, the imported ethanol is used as a blending component in gasoline, and jet fuel (considered a like commodity) is exported to qualify for the drawback. Some critics estimate that as much as 75% or more of the duties were eligible for the drawback in 2006.[28] Therefore, critics question the effectiveness of the ethanol duties and the CBI exemption.

On December 20, 2006, President Bush signed the Tax Relief and Health Care Act of 2006 (P.L. 109-432). Among other provisions, the act extended the duty on imported ethanol through December 31, 2008, but did not address the duty drawback provisions or the CBI preference.

Vehicle Purchase Requirements

The Energy Policy Act of 1992 established mandatory alternative fuel vehicle purchase requirements for various vehicle fleets.[29] Under the law, 75% of the passenger vehicles purchased by federal and state vehicle fleets must be capable of operating on alternative fuels; 90% of the vehicles purchased by alternative fuel providers must be alternative fuel vehicles.[30]

The alternative fuel vehicle provisions of EPAct 1992 have been criticized as ineffective because, while EPAct 1992 requires the purchase of vehicles capable of operating on alternative fuels, it did not mandate the use of alternative fuels. In most cases, the vehicles purchased to meet the requirement are dual-fuel vehicles (i.e., they can operate on either a conventional fuel or an alternative fuel). Those vehicles are primarily fueled using gasoline, because gasoline tends to be less expensive and more widely available than alternative fuels because the infrastructure to provide alternative fuels is limited compared with the existing infrastructure for gasoline and diesel fuel.

In addition, despite the vehicle purchase mandate, in previous years many agencies failed to meet their statutory obligation. As a result, in 2002 the Center for Biological Diversity filed a lawsuit with the U.S. District Court for the Northern District of California. In July 2002, the court ruled that several federal agencies failed to meet their quotas and ordered those agencies to prepare reports on their compliance with EPAct, which those agencies have completed.[31] Since that time, most agencies have complied with the requirement; in FY2007, the most recent year data are available, all covered federal fleets met the requirement.[32]

The Energy Policy Act of 2005 (Section 701) modified the requirements for EPAct 1992 compliance. All dual-fuel vehicles purchased to meet the EPAct quotas are required to operate on alternative fuels, unless an agency is granted a waiver by the Secretary of Energy. However, it is unclear whether this requirement will significantly affect federal agency alternative fuel use. The Secretary of Energy is required under the law to conduct a study of the effectiveness of the EPAct requirements. Further, Section 703 of EPAct 2005 allows state and fuel provider fleets to petition the Department of Energy (DOE) to waive the vehicle purchase requirement if the fleet certifies other fuel-saving measures (e.g., using higher-efficiency conventional vehicles or hybrids).

On January 28, 2008, President Bush signed the National Defense Authorization Act for Fiscal Year 2008 (P.L. 110-181). Among other provisions, the law amends the definition of "alternative fuel vehicle." Under the new definition, fleets covered by EPAct 1992 will be granted credits for the purchase of hybrid, advanced diesel,[33] and fuel cell vehicles, in addition to those alternative fuel vehicles already allowed.

In addition to the requirements for federal, state, and fuel provider fleets, EPAct 1992 grants the DOE the authority to extend the requirements to local government and private fleets. However, as of 2002, DOE had not made a determination on requirements for local and private fleets. As part of the above lawsuit, the Center for Biological Diversity also asked the court to force DOE to promulgate new rules. In ruling on the above case, the U.S. District Court for the Northern District of California ordered DOE to establish a timeline for a new rulemaking. DOE compiled a timeline and, on March 4, 2003, it issued a rulemaking determining that such a program would not promote the goals of EPAct, neither reducing dependence on foreign oil nor leading to greater use of alternative fuel vehicles.[34]

The American Recovery and Reinvestment Act of 2009 (H.R. 1) appropriates $300 million to the General Services Administration for the purchase of vehicles with high fuel economy, including hybrid, plug-in hybrid, and pure electric vehicles.

Vehicle Purchase Tax Incentives

Some supporters of alternative fuel and advanced technology vehicles argue that tax incentives for the purchase of vehicles and fuels are more effective than any purchase mandate. In addition to the mandatory purchase requirements, EPAct 1992 established tax incentives for the purchase of electric vehicles and "clean-fuel vehicles," including alternative fuel and hybrid vehicles. The Energy Policy Act of 2005 (Section 1341) significantly expanded and extended the vehicle purchase incentives, establishing tax credits for the purchase of fuel cell, hybrid, alternative fuel, and advanced diesel vehicles. For passenger vehicles, the credit is worth as much as $3,400 for hybrids and advanced diesels, and as much as $4,000 for alternative fuel vehicles, depending on vehicle attributes. The expiration date for the incentives also varies depending on the technology.[35]

In the case of hybrid and advanced diesel vehicles, the number of vehicles eligible for the credit is limited for each vehicle manufacturer. Starting the second calendar quarter after a manufacturer sells the 60,000th vehicle eligible for the credit, the credit for that manufacturer's vehicles is reduced. Currently, only Toyota and Honda have sold enough vehicles to trigger a phaseout. For Toyota (and Lexus) hybrids purchased after September 30, 2006, the credit was reduced by 50%; the credit was reduced to 25% for vehicles purchased after March 31, 2007, and is zero for vehicles purchased after September 30, 2007. Honda's phaseout began January 1, 2008, and the credit for Honda vehicles reached zero on January 1, 2009. Other manufacturers have yet to hit the 60,000 vehicle mark.[36]

The Emergency Economic Stabilization Act of 2008 established a tax credit for the purchase of plug-in vehicles, both pure electric vehicles and plug-in hybrids (i.e., gasoline/electric hybrid vehicles that can fuel on gasoline or be recharged from the electric grid.) For passenger vehicles, the credit is a maximum of $7,500, depending on the vehicle's battery capacity. After sales of vehicles eligible for the credit exceeds a total of 250,000 for all manufacturers, the credit is phased out.

The American Recovery and Reinvestment Act of 2009 (H.R. 1) would significantly modify the plug-in credits: the bill would eliminate the credit for vehicles above 14,000 pounds; establish a credit of up to $2,500 for 2-wheeled, 3-wheeled, and low-speed 4-wheeled plug-in vehicles; and establish a credit of up to $4,000 for the conversion of existing vehicles to run on battery power. The bill would also allow purchasers to claim the plug-in, alternative fuel vehicle, and hybrid tax credits even if they are subject to the

Alternative Minimum Tax (AMT)—currently, taxpayers subject to the AMT may not claim these credits.

Biodiesel

Biodiesel is a synthetic diesel fuel produced from oils, including soybean and canola oils, animal fats, and recycled cooking grease.[37] It can be blended with conventional diesel fuel and used in diesel engines with few or no modifications. Further, with some engine modifications, it can be used in a nearly pure form. Because biodiesel can displace conventional diesel without the use of new (and in many cases costly) vehicles, there is growing interest in its use. Further, because it can be produced from agricultural products, farmers (especially soybean and canola farmers) and some environmentalists have a keen interest in its development as a way to promote rural economies, reduce agricultural wastes, and limit greenhouse gas emissions. However, biodiesel production is currently expensive: wholesale biodiesel from virgin oils can cost up to two times more than conventional No. 2 diesel; biodiesel from recycled grease is less expensive but still costs considerably more than conventional diesel.

The cost barriers for biodiesel production have generated interest in providing tax incentives for biodiesel, in the form of either a production tax credit or an excise tax exemption, or both. Further there is interest in developing new technologies to help reduce production costs. However, the organic oils used as raw materials are one of the largest costs in production. Therefore, to significantly reduce biodiesel production costs, the costs of soybean oil and other oils would need to decrease substantially.

As was stated above, the American Jobs Creation Act, as amended, provides a tax credit of up to $1.00 per gallon for the sale and use of biodiesel. In addition, the law provides an excise tax credit for biodiesel blends (i.e., biodiesel and conventional diesel). Producers are eligible for one credit or the other, but not both (see "Fuel Tax Incentives" above). These credits were set to expire at the end of 2006; the Energy Policy Act of 2005 (P.L. 109-58) extended these credits through 2008. Further, EPAct 2005 established a credit of $0.10 per gallon for small agribiodiesel producers, and a $1.00-per-gallon credit for "renewable diesel"—diesel fuel produced from biomass through a different process than the biodiesel production process. The Emergency Economic Stabilization Act (P.L. 110-343) further extended these credits through 2009. While these tax credits generally do not make biomass-based diesel fuels less expensive than conventional diesel, they do help make them more cost-competitive.

Hydrogen and Fuel Cells

Over the past few years, interest has grown substantially in hydrogen fuel and fuel cells. Hydrogen fuel can be produced using any energy source, and has thus been touted as a way to limit dependence on energy imports. Further, when hydrogen is used in a fuel cell (a device that produces electricity by converting hydrogen to water), mostly heat and water are produced, drastically reducing or eliminating vehicle emissions. However, hydrogen fuel production is currently very expensive, as are fuel cells. In addition, depending on the original fuel source, overall fuel-cycle emissions can be a key concern.[38]

Because of the potential benefits from hydrogen and fuel cells, and because of the existing technical and cost barriers to their commercialization, the Bush Administration strongly supported research and development (R&D). In January 2002, the Bush Administration announced the FreedomCAR initiative, which promotes cooperative R&D between the "Big Three" American auto manufacturers (Chrysler, Ford, and General Motors) and the federal government. While the partnership is conducting research on many automotive technologies, hydrogen and fuel cell vehicles are a key focus. Further, in his January 2003 State of the Union address, President Bush announced the Hydrogen Fuel Initiative, which increased federal spending on hydrogen fuel and stationary fuel cell R&D. Overall, the President requested $1.8 billion between FY2004 and FY2008 for both initiatives, including a $720 million increase in funding from earlier appropriations. Over that time, Congress appropriated a total of $1.4 billion for the initiatives.[39] The Energy Policy Act of 2005 authorizes a total of $3.3 billion through FY20 10 for fuel cell and hydrogen R&D.[40]

Opponents of the initiatives argue that hydrogen fuel and fuel cells may never be commercialized and that the initiatives draw funding away from near-term technologies such as hybrid vehicles. Further, some argue that research and development alone will not reduce petroleum dependence and that Congress should instead consider tightening fuel economy standards for all vehicles. As noted earlier, Congress did tighten fuel economy standards for all vehicles in the Energy Independence and Security Act of 2007 (P.L. 110-140).

ALTERNATIVE FUEL AND ADVANCED VEHICLE TECHNOLOGY PROVISIONS IN THE AMERICAN RECOVERY AND REINVESTMENT ACT OF 2009

Table 1. Comparison of Alternative Fuel and Advanced Vehicle Technology Provisions in H.R. 1

Topic	House Version	Senate Version	Conference Version	Comments
Advanced Battery Manufacturing	Appropriates $1 billion for loan guarantees for advanced battery manufacturing, and $1 billion for grants for advanced battery manufacturing facilities under Secs. 135 and 136 of EISA, respectively	Appropriates $2 billion for grants for advanced battery manufacturers	Similar to Senate provision	The House version makes direct reference to authorizations under EISA, while the Senate and Conference versions do not
Grants to States for Transportation Electrification	Appropriates $200 million for transportation electrification projects under Sec. 131 of EISA	Appropriates $200 million for a transportation electrification[a]	Appropriates $400 million for transportation electrification[b]	Sec. 131 of EISA authorizes electrification grants for a variety of transportation modes, including highway vehicles, airport ground support vehicles, and ships
Grants for Advanced Vehicles	Appropriates $400 million for grants through the Clean Cities program for the purchase of alternative fuel and advanced technology vehicles under Sec. 721 of EPAct 2005	Appropriates $350 million through the Clean Cities program for the purchase of alternative fuel and fuel cell vehicles[a]	Similar to House provision, but appropriates $300 million for Sec. 721 of EPAct 2005[b]	Sec. 721 of EPAct 2005 authorizes grants to states, localities, and metropolitan transit agencies for the purchase of alternative fuel and advanced technology vehicles
GSA Purchases of Fuel Efficient and Alternative Fuel Vehicles	Appropriates $600 million for the purchase of alternative fuel and fuel efficient vehicles for the federal fleet	Appropriates $300 million for the purchase of vehicles with higher fuel economy (including advanced technology vehicles)	Similar to Senate provision	
Diesel Emissions Reduction	Appropriates $300 million for diesel emission reduction grants under Title VII, Subtitle G of EPAct 2005	Similar to House provision	Similar to House provision	EPA funding for this program in recent years has been around $50 million annually

Table 1. (Continued)

Topic	House Version	Senate Version	Conference Version	Comments
Expansion of Alternative Fuel Refueling Infrastructure Tax Credit	For 2009 and 2010, expands the current credit for the installation of alternative fuel refueling infrastructure: increases percentage credit for retail installations to 50% and maximum credit to $50,000; for hydrogen retail infrastructure, maintains 30% credit but increases maximum to $200,000; increases residential credit to $2,000	Similar to House provision, but expands credit to include electric vehicle recharging stations, if certain conditions are met	Similar to House provision	Sec. 1342 of the EPAct 2005 established a tax credit for the installation of infrastructure to deliver alternative fuels—defined as ethanol, natural gas, liquefied petroleum gas, hydrogen, or fuels containing at least 20% biodiesel; the credit is 30% of the installation cost, up to $30,000 for retail installations, or up to $1,000 for residential installations
Modification of Tax Credit for the Purchase of Plug-in Vehicles	No provision	Modifies the existing tax credit for plug-in vehicles to allow a 10% credit, up to $4,000 for the purchase of 4-wheeled low-speed electric vehicles, as well as 2- and 3-wheeled electric vehicles; establishes a 10% credit, up to $4,000 for the conversion of an existing vehicle to battery power	Modifies the existing tax credit to cap the per-vehicle credit at $7,500 for light-duty vehicles and heavy-duty vehicles up to 14,000 pounds gross weight; replaces the 250,000 total vehicle limit for phase-out of the credit with a 200,000 per manufacturer limit; eliminates the credit for heavier vehicles; establishes a credit of up to $2,500 for low-speed 4-wheel vehicles, as well as 2- and 3-wheeled electric vehicles; establishes a credit of up to $4,000 for the conversion of an existing vehicle to battery power; allows plug-in credit (as well as other alternative fuel and	Section 205 of EESA established a tax credit for the purchase of a new plug-in vehicle; the credit is based on the battery capacity of the vehicle, and is capped at $7,500 for light-duty vehicles and up to $15,000 for the heaviest vehicles; when total U.S. sales of vehicles eligible for the credit reaches 250,000, the credit begins to phase out; currently purchasers may not claim the plug-in credit and related credits for alternative fuel and advanced technology vehicles if they are subject to the AMT

Topic	House Version	Senate Version	Conference Version	Comments
			advanced vehicle credits) as a personal credit against the Alternative Minimum Tax (AMT)	
Biorefinery Grants and Loan Guarantees	No Provision	Appropriates $200 million for 2008 Farm Bill Biorefinery Assistance Program	No provision	As yet unfunded, this program would provide financial assistance for the development, construction, and conversion of plants to produce advanced biofuels

Source: CRS Analysis of: H.R. 1 as passed by the House and Senate; Senate Committee on Appropriations, *Appropriations Provisions of the American Recovery and Reinvestment Act*, February 6, 2009; *Joint Explanatory Statement of the Committee of Conference on H.R. 1*; H.Rept. 111-16.

a. Senate report language specifies this appropriation.

b. Conference report language (Joint Explanatory Statement) specifies this appropriation

Hybrid Vehicles

Hybrid gasoline/electric (and diesel/electric) vehicles are becoming increasingly popular in the United States. Hybrids combine a gasoline (or diesel) engine with an electrical motor system to improve efficiency.[41] If their use becomes more widespread, they could help improve the overall efficiency of the vehicle fleet and could help limit oil consumption. Further, they could do so without significant changes to existing infrastructure, which has been a key barrier to the expanded use of alternative fuel vehicles. By January 2009, Ford, General Motors, Honda, Nissan, Mazda, and Toyota offered vehicles with hybrid powertrains. At the present time, only hybrid passenger cars, SUVs, and pickups are available in the United States, but hybrid versions of other vehicle models and classes are expected in the near future.

Because of their energy and environmental benefits, some states have provided drivers of hybrid vehicles an exemption from high occupancy vehicle (HOV) lane requirements. Under TEA-21 (which expired on September 30, 2003), states had the authority to grant HOV exemptions for so- called "Inherently Low Emission Vehicles" (ILEVs). The ILEV standard requires that a vehicle have no evaporative emissions, a standard that is not met by any current hybrid. However, because of the reduced emissions and improved fuel economy of hybrid vehicles, there has been congressional interest in explicitly granting states the right to exempt them from HOV lane requirements. While not addressing hybrids directly, the final version of the 2005 surface transportation reauthorization act (P.L. 109-59) permits states to exempt certain high-efficiency vehicles from HOV restrictions.

Further, as was stated above, the Energy Policy Act of 2005 and the Emergency Economic Stabilization Act of 2008 expanded the incentives for the purchase of hybrid and plug-in hybrid vehicles (see "Vehicle Purchase Tax Incentives" above).

FOR ADDITIONAL READING

California Energy Commission. *ABCs of AFVs: A Guide to Alternative Fuel Vehicles.* Sacramento, CA. November 1999.

Electric Drive Transportation Association. *Technology/Vehicle Overview.* Washington, DC. Updated January 2009. http://www.electricdrive.org

Fuel Cells 2000. *Online Fuel Cell Information Center.* Washington, DC. Updated January 2009. http://www.fuelcells.org/

Methanol Institute. *Methanol Institute Homepage.* Washington, DC. Updated January 2009. http://www.methanol.org/

National Biodiesel Board. *Biodiesel Basics.* Jefferson City, MO. Updated January 2009. http://www.biodiesel

National Hydrogen Association. *General Information.* Washington, DC. Updated January 2009. http://www.hydrogenassociation.org/general/index.asp

Natural Gas Vehicle Coalition. *About NGVs.* Washington, DC. Updated November 2008. http://www.ngvc.org/about_ngv/index.html

Propane Education and Research Council. *Propane Vehicle Fleets.* Washington, DC. Updated January 2009. http://www.propanecouncil.org/trade

Renewable Fuels Association. *Ethanol Industry Outlook 2008.* Washington, DC. February 2008. http://www.ethanolrfa.org/industry

U.S. Department of Energy, Clean Cities Program. *Alternative Fuels Data Center.* Washington, DC. Updated October 2008. http://www.eere.energy

U.S. Department of Energy. *Fuel Cell Report to Congress.* Washington, DC. February 2003. ——. *National Hydrogen Energy Roadmap.* Washington, DC. November 2002.

U.S. Government Accountability Office (formerly Government Accounting Office). Energy Policy Act of 1992: Limited Progress in Acquiring Alternative Fuel Vehicles and Reaching Fuel Goals. Washington, DC. February 2000. RCED-00-59.

——. *Tax Incentives for Petroleum and Ethanol Fuels.* Washington, DC. September 2000. RCED-00-301R.

U.S. Internal Revenue Service. *Hybrid Cars and Alternative Fuel Vehicles.* Updated October 3, 2008. *http://www.irs.gov/newsroom/article/0,, id*

End Notes

[1] For more information on petroleum supply and prices, see CRS Report RL32530, *World Oil Demand and its Effect on Oil Prices*, by Robert Pirog. For more information on legislative proposals to help mitigate high gasoline prices, see CRS Report RL33521, *Gasoline Prices: Causes of Volatility and Congressional Response*, by Carl E. Behrens and Carol Glover.

[2] Alternative fuels are fuels produced from sources other than petroleum, including natural gas, coal-derived fuels, agriculture-based ethanol and biodiesel, and hydrogen.

[3] Advanced technology vehicles are vehicles that use technologies other than (or in addition to) an internal combustion engine, including electric vehicles, fuel cell vehicles, and hybrids.

[4] Tax credits were established in 2005 (by the Energy Policy Act of 2005—P.L. 109-58) for the purchase of hybrid, lean-burn, alternative fuel, and fuel cell vehicles.

[5] For more information on EISA, see CRS Report RL34294, *Energy Independence and Security Act of 2007: A Summary of Major Provisions*, by Fred Sissine.

[6] For background on alternative fuels, including legislative history, see CRS Report RL30758, *Alternative Transportation Fuels and Vehicles: Energy, Environment, and Development Issues*, by Brent D. Yacobucci. For background on advanced vehicle technologies, seeCRS Report RL30484, *Advanced Vehicle Technologies: Energy, Environment, and Development Issues*, by Brent D. Yacobucci.

[7] Alternative fuel providers are businesses that sell or distribute alternative fuels.

[8] For more information on fuel economy standards, see CRS Report RL33413, *Automobile and Light Truck Fuel Economy: The CAFE Standards*, by Brent D. Yacobucci and Robert Bamberger.

[9] Through 2008, the tax credit was valued at $0.51 per gallon. The 2008 Farm Bill lowered the credit to $0.45 per gallon in the first year after U.S. ethanol supply exceeds 7.5 billion gallons. Through October 2008, annual U.S. ethanol consumption had already exceeded 7.8 billion gallons: Renewable Fuels Association, *The Industry - Statistics*, 2008 Monthly U.S. Fuel Ethanol Production/Demand, Washington, DC, http://www.ethanolrfa.org/industry

[10] Through 2008, the credit for biodiesel produced from recycled materials was $0.50 per gallon. EESA eliminated the distinction between biodiesel fuels produced from different feedstocks.

[11] For more information on tax and non-tax incentives for ethanol and biodiesel, see CRS Report R40110, *Biofuels Incentives: A Summary of Federal Programs*, by Brent D. Yacobucci.

[12] U.S. Internal Revenue Service, Notice 2007-37: Renewable Diesel, April 23, 2007.

[13] P.L. P.L. 110-343, Division B, Sec. 202.

[14] For additional background on the MTBE issue, see CRS Report RL32787, *MTBE in Gasoline: Clean Air and Drinking Water Issues*, by James E. McCarthy and Mary Tiemann. For information on ethanol, see CRS Report RL33290, *Fuel Ethanol: Background and Public Policy Issues*, by Brent D. Yacobucci.

[15] As classified under the old 1-hour ozone standard that was replaced with a new, 8-hour standard in 2004.

[16] U.S. Environmental Protection Agency (EPA), Office of Transportation Air Quality (OTAQ), Staff White Paper: Study of Unique Gasoline Blends ("Boutique Fuels"), Effects on Fuel Supply and Distribution and Potential Improvements, October 2001.

[17] In the case of MTBE, this equates to roughly 11% by volume.

[18] Renewable Fuels Association, "New Jersey Bans MTBE," Ethanol Report, Issue #226, July 15, 2005.

[19] USDA estimates the in 2008/2009, that percentage will increase to 34%. U.S. Department of Agriculture, Economic Research Service, *Feed Outlook*, January 14, 2009.

[20] For a detailed comparison of the renewable fuels legislation, see CRS Report RL32865, *Renewable Fuels and MTBE: A Comparison of Provisions in the Energy Policy Act of 2005 (P.L. 1 09-58 and H.R. 6)*, by Brent D. Yacobucci, Mary Tiemann, and James E. McCarthy.

[21] While the original mandate in P.L. 109-5 8 covered only gasoline, the expanded mandate applies to all transportation fuels.

[22] For more information on cellulosic biofuels, see CRS Report RL34738, *Cellulosic Biofuels: Analysis of Policy Issues for Congress*, by Tom Capehart.

[23] A mixture of hydrogen and carbon monoxide that can be used to produce a variety of chemicals and fuels.

[24] On December 23, 2008, the Department of Energy published a proposed rule to establish the incentive program. U.S. Department of Energy, "Notice of Proposed Rulemaking: Production Incentives for Cellulosic Biofuels; Reverse Auction ," 73 *Federal Register* 78663-78670, December 23, 2008.

[25] For more information on ethanol imports from CBI countries, see CRS Report RS21930, *Ethanol Imports and the Caribbean Basin Initiative (CBI)*, by Brent D. Yacobucci.

[26] For more information on drawbacks, see U.S. Customs Service, Drawback: A Refund for Certain Exports, Washington, February 2002.

[27] 19 U.S.C. 1313(p).

[28] Peter Rhode, "Senate Finance May Take Up Drawback Loophole As Part Of Energy Bill," EnergyWashington Week, April 18, 2007.

[29] For purposes of compliance with EPAct 1992, a covered vehicle fleet is one operated by an agency or company in a metropolitan area with at least 20 passenger vehicles in one location.

[30] For more information on vehicle purchase requirements, see CRS Report RL30758, *Alternative Transportation Fuels and Vehicles: Energy, Environment, and Development Issues*, by Brent D. Yacobucci.

[31] Center for Biological Diversity v. Abraham, N.D. Cal., No. CV-00027.

[32] U.S. Department of Energy, Office of Energy Efficiency and Renewable Energy, *Federal Fleet Compliance with EPACT and E.O. 13149/E..O. 13423,* Fiscal Years 2000 through 2007.

[33] Light-duty diesel vehicles that meet specified emissions standards.

[34] 68 *Federal Register* 10319.

[35] For more information on vehicle tax incentives, see CRS Report RS2235 1, *Tax Incentives for Alternative Fuel and Advanced Technology Vehicles*, by Brent D. Yacobucci.

[36] For more information on the hybrid vehicle tax credit, see Internal Revenue Service, *Summary of the Credit for Qualified Hybrid Vehicles*, at [http://www.irs.gov/newsroom/article/0,,id updated October 3, 2008.

[37] For more information on biodiesel, see CRS Report RL32712, *Agriculture-Based Renewable Energy Production*, by Randy Schnepf, and CRS Report RL3 0758, *Alternative Transportation Fuels and Vehicles: Energy, Environment, and Development Issues*, by Brent D. Yacobucci.

[38] For example, depending on the technology used, processing coal into hydrogen could lead to significantly higher emissions of toxic compounds and carbon dioxide.

[39] Congress agreed to increase funding for hydrogen and fuel cell research from $185 million in FY2003 to $266 million in FY2004, $305 million in FY2005, $335 million in FY2006, $335 million for FY2007, and approximately $400 million for FY2008.

[40] For more information on the Bush Administration's initiatives, see CRS Report RS2 1442, *Hydrogen and Fuel Cell Vehicle R&D: FreedomCAR and the President's Hydrogen Fuel Initiative*, by Brent D. Yacobucci.

[41] For more information on hybrid vehicles, see CRS Report RL3 0484, *Advanced Vehicle Technologies: Energy, Environment, and Development Issues*, by Brent D. Yacobucci.

In: U.S. Energy and the Environment… ISBN: 978-1-61668-017-6
Editors: Roland H. Terrison pp. 117-177

Chapter 6

CARBON TAX AND GREENHOUSE GAS CONTROL: OPTIONS AND CONSIDERATIONS FOR CONGRESS

Jonathan L. Ramseur and Larry Parker

SUMMARY

Market-based mechanisms that limit greenhouse gas (GHG) emissions can be divided into two types: quantity control (e.g., cap-and-trade) and price control (e.g., carbon tax or fee). To some extent, a carbon tax and a cap-and-trade program would produce similar effects: Both are estimated to increase the price of fossil fuels, which would ultimately be borne by consumers, particularly households. Although there are multiple tools available to policymakers that could control GHG emissions—including existing statutory authorities—this chapter focuses on a carbon tax approach and how it compares to its more frequently discussed counterpart: cap-and-trade.

If policymakers had perfect information regarding the market, either a price (carbon tax) or quantity control (cap-and-trade system) instrument could be designed to achieve the same outcome. Because this market ideal does not exist, preference for a carbon tax or a cap-and-trade program ultimately depends on which variable one wants to control—emissions or costs. Although there are several design mechanisms that could blur the distinction, the gap between price control and quantity control can never be completely overcome.

A carbon tax has several potential advantages. With a fixed price ceiling on emissions (or their inputs—e.g., fossil fuels), a tax approach would not cause additional volatility in energy prices. A set price would provide industry with better information to guide investment decisions: e.g., efficiency improvements, equipment upgrades. Economists often highlight a relative economic efficiency advantage of a carbon tax, but this potential advantage rests on assumptions—about the expected costs and benefits of climate change mitigation—that are uncertain and controversial. Some contend that a carbon tax may provide implementation advantages: greater transparency, reduced administrative burden, and relative ease of modification.

The primary disadvantage of a carbon tax is that it would yield uncertain emission control. Some argue that the potential for irreversible climate change impacts necessitates the emissions certainty that is only available with a quantity-based instrument (e.g., cap-and-trade). Although it may present implementation challenges, policymakers could devise a tax program that allows some short-term emission fluctuations, while progressing toward a long-term emission reduction objective. Proponents argue that short-term emission fluctuations would be preferable to the price volatility that might be expected with a cap-and-trade system.

Although a carbon tax could possibly face more political obstacles than a cap-and-trade program, some of these obstacles may be based on misunderstandings of the differences between the two approaches or on assumptions that the tax would be set too low to be effective. Carbon tax proponents could possibly address these issues to some degree, but there remains considerable political momentum for a cap-and-trade program.

INTRODUCTION

A variety of efforts that seek to reduce greenhouse gas emissions (GHG)[1] are currently underway or being developed on the international, national, and sub-national level (e.g., individual state actions or regional partnerships). One option (of many, see text box below—"Other Policy Options for Addressing GHG Emissions") for controlling GHG emissions is to apply a tax or fee on GHG emissions or the inputs that create them. This type of approach is commonly called (and referred to in this chapter as) a carbon tax,[2] whether it would apply to CO_2 emissions alone or to multiple GHGs, including some that may have no molecular carbon.[3] This chapter does not provide a comprehensive comparison and analysis of the multiple policy tools available to Congress that would address climate change. Instead, this chapter focuses on the policy considerations of using of a carbon tax to control GHG emissions.

Governments may impose taxes for a variety of purposes. The primary reason that governments impose taxes is to raise revenue to fund various objectives or services: e.g., national defense, public education, social security, etc. Generally, governments raise these revenue streams by placing a tax on activities that are recognized as desirable ("economic goods") such as income, employment, and investment. While this tax placement ensures a relatively steady flow of revenue (often the primary objective of the tax), economists generally describe such taxes as distortionary, because the taxes discourage the "good" activity. For example, many economists have argued that payroll and income taxes discourage employment and investment.[4] If these taxes were reduced, the incentives to increase labor and investment would be greater.

Economists maintain that levying a charge on pollution (sometimes referred to as a Pigouvian tax)[5] would be an efficient way to correct an inherent failure in a particular market. A basic economics principle is that market prices may not reflect the social cost of resource use (e.g., fossil fuel combustion) when economic activities result in pollution (e.g., CO2 emissions). If social costs are not included, the market price of the resources will not reflect their true costs. For example, in terms of climate change policy, the price of using fossil fuels, particularly coal, does not reflect the costs—i.e., climate change-related damages—associated with CO_2 emissions.

In economics parlance, the social cost not reflected in the market price is called an "externality." A pollution tax would internalize the external costs by making the party who profits from the polluting activity include the external costs in the price of the good or service. Policymakers could place a pollution tax on GHG emissions or the inputs that create them. By attaching a price to GHG emissions, a carbon tax would stimulate GHG emission reduction. If the tax were placed on emissions, entities directly subject to the tax, such as power plants, would have an incentive to take actions—e.g., energy efficiency improvements or equipment upgrades—to lower tax payments. If the tax were placed on emission inputs—e.g., fossil fuels—the price of carbon- intensive energy sources, primarily coal, would increase relative to low-carbon fuels (**Table A-1** of this chapter—located in the **Appendix**—includes estimates of price increases to fossil fuels and motor gasoline based on different carbon tax rates). Energy consumers—e.g., power plants, industry, households, etc.—would be encouraged to (1) switch to less carbon-intensive fuels; (2) use less energy or use energy more efficiently; and (3) prefer products or services that are lower- priced by virtue of incorporating less emission tax. Each of these activities would reduce GHG emissions compared to a business-as-usual track.

These expected behavioral changes mirror the activities that are forecast for a potential cap-andtrade program. Both a carbon tax and a cap-and-trade system would place a price on carbon. Both a carbon tax and cap-and-trade system are intended (and expected) to increase the price of coal, oil, and natural gas. Under either program, these price increases would ultimately be borne by energy consumers, both businesses and households. These price increases are integral to a market-based approach to GHG emission reduction, because they send more accurate information to purchasers about the full cost of their choices.

This chapter begins with an overview of the fundamental choices involved between a cost (tax) and a quantity (cap) control instrument. This includes a discussion of policy tools that could be employed to bridge the gap between a carbon tax and a cap-and-trade program. Following this overview, the report analyzes the potential advantages and disadvantages of a carbon tax. In many cases, carbon tax attributes are compared with those of a cap-and-trade program. The next section discusses implementation issues for a carbon tax, including where to apply the tax, at what level to set the tax, and options for distributing the tax revenues. The final section provides conclusions.

Other Policy Options for Addressing GHG Emissions

For policymakers considering actions to address climate change, a variety of policy instruments is available. Although current attention has largely focused on market-based mechanisms, primarily cap-and-trade systems, non-market policy tools may be the most practical option to address some emission sources. Moreover, Congress may consider complementary approaches to market mechanisms to improve their effectiveness. In particular, many experts maintain that the GHG emission targets specified in recent legislation would require development and deployment of improved (low-carbon) technologies. To further this effort, some argue that Congress should address technology stimulation directly or as a supplement to the primary climate change mitigation policy. In addition, Congress has already addressed some specific sectors (cars and light trucks, and government buildings) through performance standards.

These standards may be further strengthened, independent of climate change legislation.

Efforts are underway to address GHG emissions using the existing Clean Air Act (CAA) statute. The Environmental Protection Agency (EPA) has received at least eight petitions asking it to use its CAA authority to regulate GHG emissions from multiple categories of mobile sources. The agency faces lawsuits seeking GHG regulations (per CAA authority) from power plants, refineries, and other stationary sources. In 2007, the Supreme Court found (Massachusetts v. EPA) that GHG emissions are air pollutants under the CAA, and that EPA has the authority to promulgate GHG emission controls.

For more information, see CRS Report R40145, *Clean Air Issues in the 111th Congress*, by James E. McCarthy; and *CRS Report RL34513, Climate Change: Current Issues and Policy Tools*, by Jane A. Leggett.

COST OR QUANTITY CONTROL: AN OVERVIEW

If policymakers choose to establish a market-based mechanism to control GHG emissions, a fundamental decision would be whether to use a price instrument, such as a carbon tax, or a quantity instrument, such as an emissions cap.[6]

Economic Theory vs. Uncertainty

In an economically efficient market with perfect information, either a price (carbon tax) or quantity control instrument (cap-and-trade system) could be designed to achieve the same outcome.[7] **Figure 1** illustrates this basic economic principle. The intersection of marginal costs[8] and marginal benefits[9] would provide the point (of economic efficiency) at which to set the price or quantity limit. At this point in the figure, the abatement costs[10] equal benefits received from abatement. Per economic theory, emission abatement above or below this point would not be economically efficient.[11] The dashed lines indicate the efficient price (tax) and quantity (cap) limits. This figure illustrates that (with perfect information) if either a tax or cap is selected, both the emission abatement level and cost of abatement would be identical. For instance, if a cap were chosen, covered sources (e.g., power plants) would abate emissions until they reach the cap of Q*, at which point the marginal cost of abatement would equal P*. If a tax were chosen, sources would abate emissions until the marginal cost of abatement reached the tax level (P*), at which point the emission abatement level equals Q*. In either case, total abatement is Q*, and the total cost of the program is the shaded area under the marginal abatement cost curve.

However, the illustration in **Figure 1** depicts economic theory, not economic reality: An economically efficient market with perfect information does not exist. The primary knowledge gaps are the shapes (i.e., slope) and levels (i.e., magnitude) of the marginal abatement cost and marginal benefit curves. These benefit and cost curves are influenced by multiple variables, including uncertain future emission baselines. Baseline estimates are related to projected population levels and economic growth,[12] factors that are also unknown and difficult to predict.[13] Another uncertain factor, which will have a substantial influence over mitigation costs, is the development and deployment of low-carbon technology.

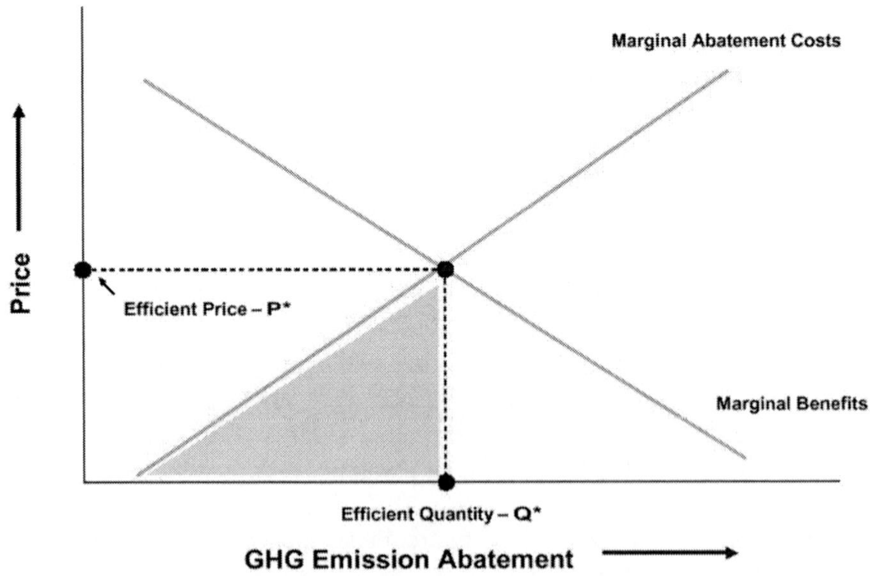

Source: Prepared by CRS.

Figure 1. Illustration of Price Versus Quantity

Without accurate information regarding marginal costs and benefits, policymakers cannot know the efficient quantity or price from which to establish an emissions cap or carbon tax. Thus, in one sense, preference for a price (carbon tax) or a quantity limit (emissions cap) depends on one's preference for uncertainty—either uncertain emissions or uncertain program costs. A tax would set a ceiling on the marginal price of emissions (and thus the total cost of the program), but emissions would be uncertain. A cap would hold emissions to a set level, but the price of emission reductions (and thus program costs) would be uncertain.[14] However, as discussed below, the choice is not as simple as it may seem.

A Stark Choice or a Policy Continuum?

There are multiple design elements that can be included with a cap-and-trade program that blur the distinction between price and quantity control. Similarly, a carbon tax program could include flexible design mechanisms allowing policymakers to alter the tax rate, if they determine that emission reductions are not proceeding at a desirable pace. With these design elements available, Congress is presented with a policy continuum, rather than a stark policy dichotomy.

A flexible emissions cap

Congress could enact a cap-and-trade program that includes mechanisms that provide flexibility in terms of controlling costs. **Figure 2** illustrates the options available. As the design elements in **Figure 2** (the "bridge pieces") are added to a cap-and-trade program, the program more closely resembles a carbon tax. These elements are briefly described below:

- Banking: Covered sources may save (bank) emission allowances[15] and submit them in future years. Banking reduces the absolute cost of compliance by making annual emissions caps flexible over time. In addition, banking reduces price volatility to some degree.[16]
- Borrowing: Covered sources may borrow (typically at a relatively high interest rate) emission allowances allotted to future years. The ability to borrow allowances from future years should help control price spikes.

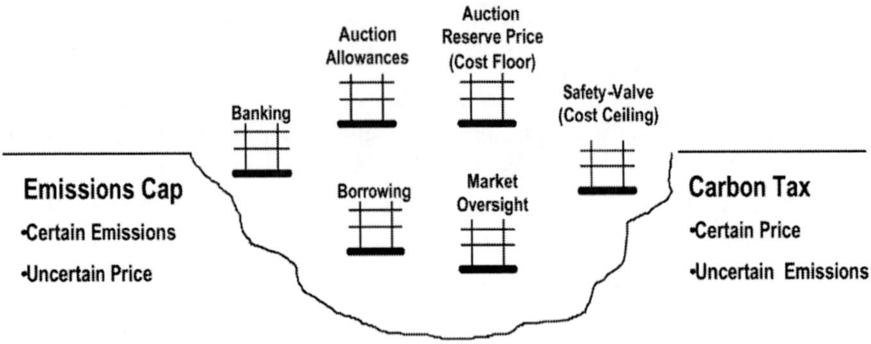

Source: Prepared by CRS.

Figure 2. Bridging the Gap between Price and Quantity Control

- Market oversight: Congress could create a regulatory mechanism (e.g., an oversight board) to observe the allowance market and implement cost-relief measures if necessary. Obviously, the authority granted to the delegated entity would determine its ability to affect program costs.
- Safety valve: Generally triggered by prices in the allowance markets, safety valves may include (1) a set price alternative to making reductions or buying allowances at the market price, (2) a slowdown in tightening the emissions cap, and (3) lengthening of the time allowed for compliance. The first type would move the program closest to price control by setting a ceiling on allowance prices and limiting their volatility.
- Auctioning allowances: Auctions may be used to allocate some or all of the emission allowances. Like a carbon tax, auctions would raise revenues to support various objectives. In addition, auctions would support price discovery.[17]
- Reserve price: A reserve price is a price below which the seller refuses to part with the item for sale. A reserve price would all but guarantee a revenue stream in an emissions allowance auction, thus operating much like a minimum tax or price floor.

A flexible carbon tax

Policymakers could establish a carbon tax that includes mechanisms for addressing long-term emission uncertainty. For example, emission sources would be required to submit emission data periodically, so that policymakers could assess the performance of the carbon tax. If Congress determined that emission reduction was inadequate, the tax rate could be adjusted to achieve a desired emissions level. The decision authority could be retained by Congress or delegated to an existing federal agency—e.g., Department of Treasury or Environmental Protection Agency—or another entity (e.g., a newly created independent board) in order to meet pre-determined emission reduction objectives.[18]

Considering the consequences of altering a carbon tax rate, Congress may be hesitant to delegate this authority. In addition, some may question whether the federal government is nimble enough to modify the carbon tax rate on a periodic basis, particularly considering the potential size of the tax (**Table 1** below). Moreover, a flexible carbon tax approach may sacrifice long-term price certainty, depending on the authority of the entity to adjust the tax rate. To address this concern, Congress may consider providing tax rate increase parameters for the delegated entity.

Limits of the policy continuum

Regardless of whether policymakers employ these additional design elements, the gap between price control and quantity control can never be completely overcome: The closest resemblance would be a cap-and-trade program that employs both a price safety-valve and an emission allowance auction with a reserve price. Even if all of the above design options are part of a capand-trade program, the price of emission allowances would still fluctuate to some degree. Likewise, a more flexible tax system would still yield short-term emission fluctuations.

The policy continuum demonstrates that a comparison between a carbon tax and an emissions cap is not a straightforward exercise. Design details are critical for an appropriate comparison. Depending upon program design, cap-and-trade programs can vary greatly in their ability to control price fluctuations and total program costs. Some of the advantages or disadvantages (depending on one's perspective) of either system can be addressed through program design. Indeed, recent federal cap-and-trade proposals have included one or more of the design elements listed above.

POTENTIAL ADVANTAGES OF A CARBON TAX

This chapter examines four potential advantages of a carbon tax approach to controlling GHG emissions. First, the report compares a carbon tax to cap-and-trade approach from an economics theory perspective, analyzing the assertion (of many economists) that a carbon tax would be more economically efficient than a cap-and-trade system. Second, it discusses the merits of price stability that a carbon tax would provide. The third discussion highlights the advantage of generating carbon tax revenues and examines various revenue applications and the trade-offs that would ensue. Finally, the report identifies potential implementation advantages, primarily practical considerations, of using a carbon tax approach.

Economic Efficiency

A primary argument supporting the use of a carbon tax is that in the presence of cost uncertainty—i.e., the marginal abatement cost curve (**Figure 1**)—a tax would potentially be more economically efficient than an emissions cap. However, this argument is grounded in assumptions that are the subject of considerable debate and controversy. This section outlines the basis for the argument and provides an analysis of the underlying assumptions.

Basis for the argument

As **Figure 1** demonstrates, if the marginal cost and benefit curves are known to policymakers, either a tax or a cap instrument could be used to achieve the same result. As discussed above, both the costs and benefits of GHG emission reduction are uncertain. However, some economists argue that the *relative* slopes of the cost and benefit curves are better understood. This information—the slope of the marginal benefit curve in comparison to the slope of the marginal cost curve—is the foundation for an argument that supports a tax (price control) over an emissions cap (quantity control).

This argument was presented in Weitzman's landmark 1974 study "Prices vs. Quantities."[19] Weitzman's primary conclusion (explained in some detail below) was that price instruments (e.g., carbon tax) would be preferred when the marginal benefit curve is relatively flat; a quantity limit (e.g., emissions cap) would be preferred when the marginal cost curve is relatively flat.[20]

Intuitively, a relatively flat marginal benefits curve suggests that each additional unit of abatement provides approximately the same benefit. In contrast, a relatively steep marginal cost curve suggests that each additional unit (i.e., ton of CO2) of emission abatement entails a greater cost than the previous unit.[21] Some economists argue that the slope of the marginal benefits curve is flat relative to the marginal cost curve.[22] This notion is the subject of intense debate, but it is grounded in the scientific processes (discussed below) through which GHG emissions generate climate change impacts.

Based on this (assumed) relationship between the expected slopes of costs and benefits, a carbon tax would be more efficient than an emissions cap. **Figure 3** and **Figure 4** below illustrate this concept. **Figure 3** provides a scenario in which the marginal benefits curve is flatter than the marginal costs curve. Without precise information regarding marginal costs and benefits, policymakers would base a tax or cap on *estimates* of the marginal cost and benefit curves. These are represented in **Figure 3** as the *expected* marginal costs and benefits of GHG emission abatement.[23] The intersection of the expected marginal abatement cost and benefit curves would be the estimated economically efficient point of regulation. From this efficient point, policymakers could apply either an emissions cap or carbon tax—at Q* or P*, respectively. Either option would produce the same outcome, if the *expected* marginal cost curve matches the *actual* marginal cost curve.[24]

However, the *actual* cost curve is unlikely to match the *expected* cost curve. Therefore, some level of economic inefficiency will exist with either a tax or an emissions cap. However, the policies are expected to yield different levels of inefficiencies. For example, consider a scenario (**Figure 3**) in which *actual* marginal abatement costs are higher than *expected.* If policymakers were to apply a carbon tax (P* in the figure) directly to emissions, emission sources would abate emissions until the marginal cost of abatement equals the carbon tax. This occurs at **Point A**, at which time covered sources would pay the tax for any additional emissions. On the other hand, if policymakers employed an emissions cap (Q*), covered sources would abate emissions until **Point B**, when the cap is met.

With a tax instrument in **Figure 3**, under higher-than-expected costs, covered sources would abate less than the optimal level—the intersection of benefits and *actual* costs; with an emissions cap, sources would abate more than the optimal level. However, **Figure 3** illustrates that the difference in efficiency losses—the shaded triangles—from the two policies is substantial.

Figure 3 also illustrates the scenario of lower-than-expected marginal abatement costs. With lower-than-expected marginal abatement costs, an emissions cap would lead to abatement that is less than optimal, whereas a tax would stimulate abatement beyond the optimal level. As with higher-than-expected costs, a tax instrument would yield less inefficiency than an emissions cap. The different efficiency losses are depicted by the striped triangles in **Figure 3**.

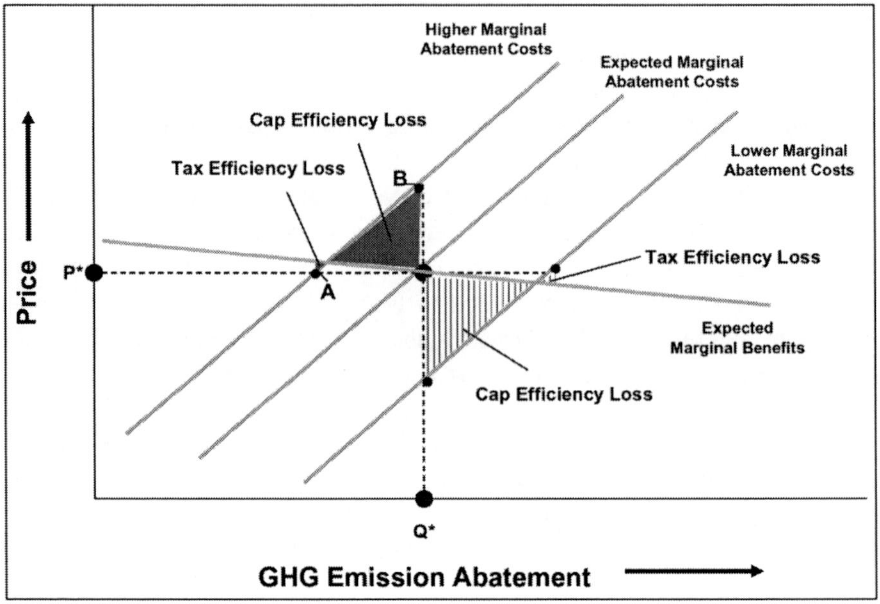

Source: Prepared by CRS.

Figure 3. Illustrative Scenario with a Relatively Flat Marginal Benefits Curve

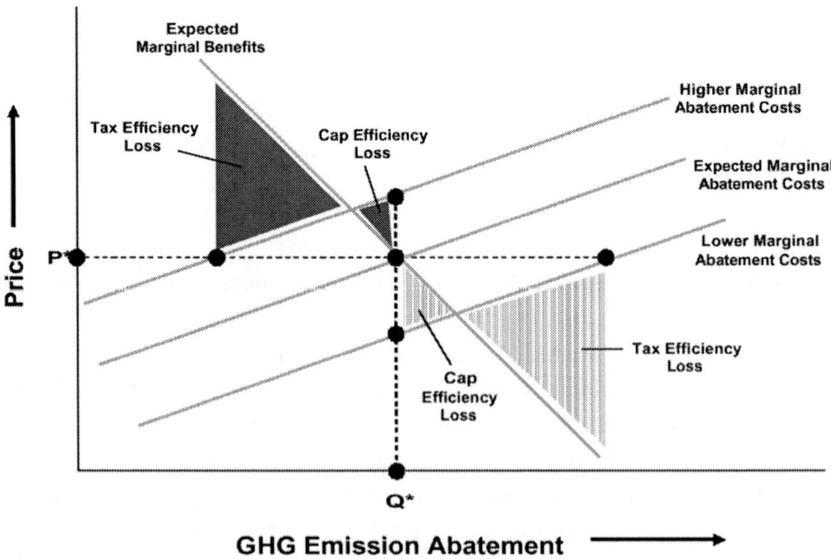

Source: Prepared by CRS.

Figure 4. Illustrative Scenario with a Relatively Steep Marginal Benefits Curve

Figure 4 illustrates the alternate scenario, in which the marginal benefits curve is relatively steep. In this situation, an emissions cap is the preferred instrument. Whether the marginal abatement costs are higher or lower than expected, the efficiency losses from a carbon tax are greater than those from an emissions cap.

Underpinnings of the argument

The argument that a carbon tax would minimize economic inefficiency in the face of uncertain costs contains several assumptions. A primary assumption concerns the relative steepness/flatness of the cost and benefit curves. In addition, there are other underpinnings of the argument that may be questioned. These issues are discussed below.

Relative flatness of benefits curve?

At the crux of the argument that a tax would maximize efficiency is the assumption that the slope of the marginal benefits curve is flat compared to the marginal costs curve (**Figure 3**). Many would assert this is the case, based on the process by which GHG emissions generate damages. In the context of climate change, the damages related to GHG emissions are due to their concentration (referred to as "stock") in the atmosphere, as opposed to their annual flow.[25] By comparison, more traditional air pollutants (e.g., sulfur dioxide, nitrogen oxide) impose damages through their annual emission flows.

With a "stock pollutant" like CO_2, the processes involved in adding or removing the gas from the atmosphere are measured, not in years, but in decades (or centuries for some gases). The current stock of CO_2 in the atmosphere is approximately 3,000 gigatons;[26] in 2004, the entire world emitted roughly 30 gigatons of CO_2. Because emission quantities would have only a minor effect on the accumulation of CO_2 in the atmosphere, the benefits from controlling the stock (or the damages from business-as-usual activities) are thought to occur slowly and steadily.[27] **Figure A-1** in the **Appendix** illustrates this concept.

However, the assertion that the benefits curve is flat assumes that there are no climatic "tipping points." Some scientists maintain that at a certain stock level—concentration of GHG emissions in the atmosphere—particularly drastic events might occur.[28] By definition, a tipping point implies that when it is reached, the course of climate change is abrupt and the marginal benefits curve becomes very steep. The climate changes may even be irreversible and catastrophic, at least for some populations and environments. Examples often cited include collapse of Greenland's ice sheet; and dieback of the Amazon rainforest.[29]

If such tipping points exist, they would dramatically alter the shape of the benefits curve, because abating emissions to avoid large adverse consequences would yield large benefits. As argued by the Stern Review: "over the long term, as the stock of GHGs grows, marginal damages are likely to rise and – as the stock reaches critical levels – marginal damages may rise sharply. In other words, the damage function is likely to be strongly convex."[30] However, the Stern Review also notes that: "to the extent that damages may relate to the *rate* of climate change, the relationship is more complex, but it remains true that the damage curve is likely to respond most to cumulative emissions over several years or even decades."[31] Depending on whether the tipping points are "several years" or "decades" away, an emissions cap may be preferred.

However, scientists and policymakers have very poor information regarding the GHG concentration levels (or rates of concentration changes) at which tipping points may occur. A 2003 economic study found that emissions caps would be "optimal only when there is a well- defined threshold with clearly catastrophic consequences on the other side."[32]

Relative steepness of the marginal cost curve?

The marginal cost curve is perhaps better understood (and less controversial) than benefits estimates. Analysis of recent cap-and-trade proposals introduced in Congress suggest that with the current portfolio of technologies the marginal cost curve does not appear flat.[33] Although the first units of emission abatement opportunities may be relatively inexpensive, the costs of abatement are likely to increase with each additional unit. However, if a carbon market spurs technology development, the marginal cost curve may flatten. This could reverse the relationship between costs and benefits, and make a cap the more economically efficient policy.[34]

Short-term time frame

In general, Weitzman's economic principle regarding the relative slopes of the marginal cost and benefit curves applies to a static or fixed time-frame. The principle is best suited to compare the costs and benefits of controlling a "flow pollutant" (such as SO2), which would yield short-term damages. However, as noted above, CO2 is a stock pollutant: its potential damages are related to total accumulation in the atmosphere; one year of emissions would likely have negligible effects. Thus, the static analysis is over-simplistic, because (by definition) it assumes that activities in different time periods are unrelated.[35] However, equipment upgrades or efficiency improvements made in one time period would affect the emissions generated in future years as well. Other time- related factors need to be considered, including the discounting of future benefits and the atmospheric decay of GHGs.

Several economic studies have addressed this issue of time, expanding Weitzman's basic principle within a dynamic framework (i.e., a multi-year time-period). In general, these studies concluded that Weitzman's basic principle carries over to a dynamic setting.[36]

Only cost uncertainty is considered

Although marginal benefit information is also uncertain, the relative-slope principle generally depends only on the marginal cost curve. The rationale behind this is that when covered sources make decisions regarding emission abatement—e.g., reduce onsite emissions or pay the carbon tax—benefits of abatement are not part of their economic calculation. Economists point out that if covered sources did consider benefits, there would be no externality in the first place.

However, in some circumstances an emissions cap may produce a more economically efficient outcome. As mentioned (in a footnote) by Weitzman (1974) and re-examined by Stavins in 1996,[37] if the uncertainty of costs and benefits are positively correlated—e.g., both the marginal benefit and cost curves are actually higher than expected—quantity control provides a more economically efficient result. This scenario is illustrated in **Figure 5**. The advantage for quantity control over price control under these circumstances depends on the degree of correlation and the magnitude of uncertainty with respect to the cost and benefits curves – the greater the difference between anticipated and actual cost and benefits values, the greater the advantage for quantity control measures. Stavins points out that this reversal in preference for price over quantity control could occur with plausible values for the two variables; however, he argues that "the advantage of price over quantities remains unless the true benefits of carbon mitigation are many orders of magnitude greater than our best estimate."[38]

Modeled efficiency gains

Several economic studies have examined the efficiency gains of using a carbon tax versus an emissions cap.

> While the analysis is more complicated and involves more than simply the relative slopes of marginal abatement and damage curves, the analyses

> consistently find that taxes dominate cap-and-trade systems for a broad range of parameter values consistent with scientific understanding of the global warming problem [emphasis added].[39]

For example, Pizer found that taxes are more efficient than an emissions cap by a factor of five to one.[40] Although this result is from 1997, similar estimates were made in more recent studies that applied different parameters.[41] The magnitude of these efficiency gains were estimated to be in hundreds of millions to several billions of dollars per year.[42]

However, a 2008 Massachusetts Institute of Technology (MIT) report points out that the modeled efficiency gains may be smaller in practice.[43] As discussed above, policymakers can design capand-trade programs with flexible components, such as banking and borrowing, that could alleviate some of the inefficiency loss. But a 2008 Resources for the Future (RFF) study indicates that banking may not help "as much as proponents might suggest." The study found that banking provisions in a cap-and-trade program would improve efficiency by 20% over a cap-and-trade without banking.[44]

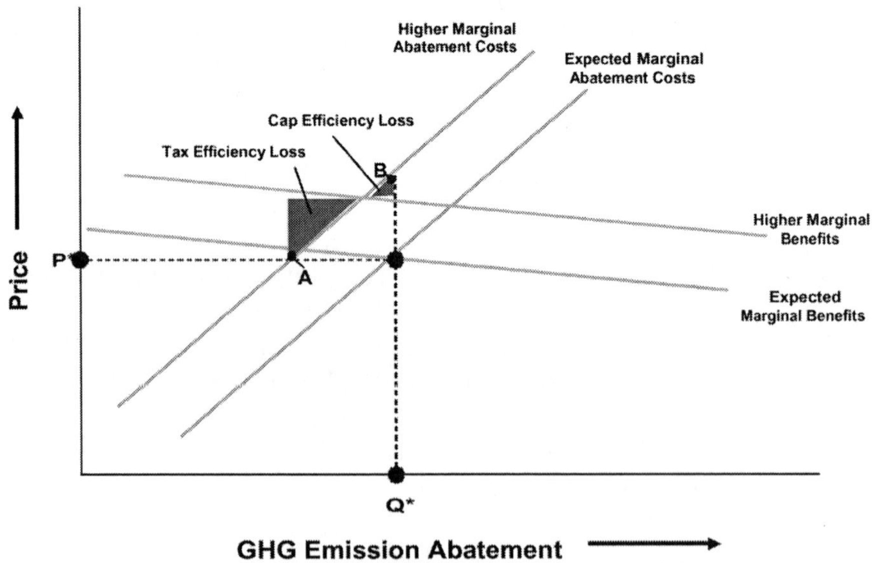

Source: Prepared by CRS.

Figure 5. Illustrative Scenario with Marginal Costs and Marginal Benefits That Are Higher Than Expected

Economic efficiency versus precaution

Despite the economic efficiency argument in favor of a carbon tax, some argue that the downside risks of tipping points—i.e., irreversible climate changes—demand that economic efficiency take a backseat to precise control of GHG reductions, achievable only with a quantity-based emissions cap. This viewpoint could be described as an application of the precautionary principle: i.e., keeping one's options open in the face of uncertainty.[45] This approach argues that, while GHG emissions are held at an exact level, scientists and economists can continue to gather information regarding abatement costs and climate change benefits. As (presumably) better information is garnered, climate change policy can be amended as necessary.

Price Stability

Depending on the chosen GHG control instrument—carbon tax or cap-and-trade—either emissions or prices would fluctuate over short-term time periods (months to years). Proponents of a carbon tax often highlight the volatile price swings that have been observed in the U.S. sulfur dioxide (SO2) emissions cap-and-trade program.[46] Between 2001 and 2006, the price of a SO2 allowance has varied by a factor of 12.[47] The inclusion and substantial use of banking has not been sufficient to overcome other factors creating price volatility, such as uncertainty regarding proposed regulatory changes in the program (e.g., The Clean Air Interstate Rule).

Although the price volatility in the SO2 market did not translate into electricity price spikes for consumers, CO2 emission allowances could play a more integral role in the U.S. economy: Fossil fuels accounted for 85% of energy consumption in 2006.[48] Nordhaus states that strict quantity limits might have "major disruptive effects on energy markets and investment planning, as well as on the distribution of income across countries, inflation rates, energy prices, and import and export values."[49] Volatile emission allowance prices may discourage investment, because firms would have less reliable cost information upon which to plan for capital improvements.[50]

A GHG cap-and-trade program may or may not have similar price swings to those experienced with the SO2 allowance market. The largest GHG cap-and-trade program currently operating is the European Union's Emission Trading System (EU ETS). During the four years of its existence, both Phase 1(2005-2007) and Phase 2 (2008-2012) EU ETS allowance prices have experienced significant volatility. Phase 1 allowance prices from the EU ETS are arguably not an appropriate comparison, because the fluctuations were largely related to program design issues.[51] Building on the experience gained during Phase 1, Phase 2 of the ETS began in 2008, and these emission allowance prices may provide a better example of the potential effects of market forces and the fluctuations that could occur under a U.S. cap-and-trade program.[52] **Figure 6** illustrates the market prices for Phase 2 allowances between 2005 and January 2009.

Tax Revenue Applications

Carbon tax proponents often cite the generation of carbon tax revenues as a primary advantage of a carbon tax. However, policymakers could design a cap-and-trade program that could deliver similar, if not identical, results. For example, if policymakers establish a cap-and-trade program and distribute 100% of the allowances through an auction process, the auction proceeds would function as tax revenues.[53] Thus the strength of this advantage depends on the design of the capand-trade program to which the tax is compared.

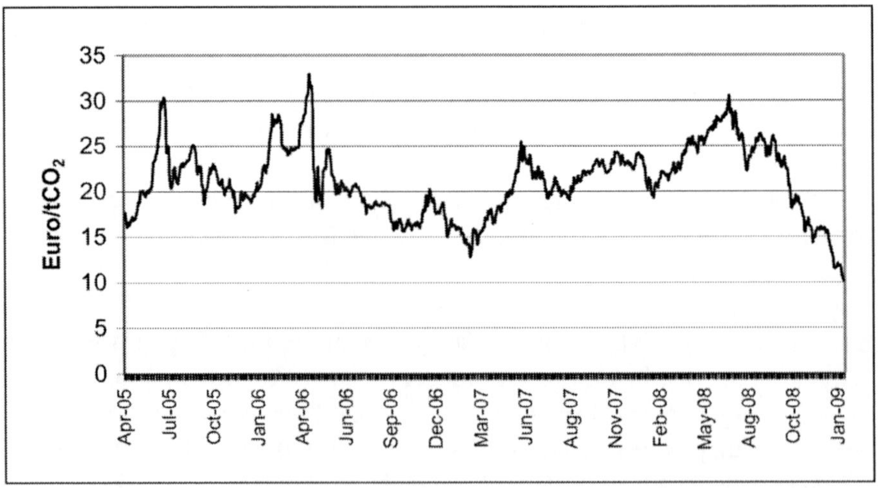

Source: Prepared by CRS with data from European Climate Exchange, at *http://www.ecxeurope.com.*

Figure 6. "Phase 2" Emission Allowance Prices in the European Union's Emission Trading System

From a public finance perspective, neither carbon tax nor auction revenues would be a reliable source of funding, because an effective carbon price (established through a tax or a cap) is expected to alter consumer behavior (e.g., demand for carbon-intensive goods). However, a carbon tax revenue stream would have a more predictable source than an allowance auction revenue stream, because the tax would be a known value, while the emission allowance price would fluctuate (as described above).

Different options for applying the tax revenue are discussed in the section "Implementation of a Carbon Tax."

Potential Implementation Advantages

Depending on the design, a carbon tax may offer several implementation advantages over a capand-trade program. These are discussed below.

Transparency

Arguably, the concept of a carbon tax is easier to explain and understand than a cap-and-trade program. A cap-and-trade system may have more moving parts, which may be challenging to follow, particularly the emission allowance allocation scheme. Moreover, as policymakers include more flexible design elements—primarily to improve efficiency and control price volatility—a cap-and-trade program would increase in complexity and potentially become less transparent. Some contend that the relative transparency of a carbon tax would help garner support for its enactment.[54] This argument may gain strength in the context of the recent financial climate, as many have grown more skeptical of seemingly complex financial structures.[55]

Although the concept of a carbon tax is arguably a simpler approach, many argue that the U.S. tax code is complex. Congress could establish a carbon tax framework that rivals the complexity of a cap-and-trade program. For instance, a carbon tax that only applies to CO2 emissions from fossil fuel combustion may be more transparent than a carbon tax that address non-CO2 GHG emissions. Moreover, policymakers could provide subsidies or exemptions to the fossil fuel industry that would run counter to a carbon tax. In addition, policymakers could allow for tax credits for carbon sequestration projects, similar to carbon offsets in a cap-and-trade regime. As with carbon offsets in a cap-and-trade program, this would require a further level of administrative responsibilities, and potentially weaken the program if the sequestration projects lack credibility.[56]

Ironically, transparency, particularly in regards to costs, could be a political liability for a carbon tax. Although both a carbon tax and a cap-and-trade program would impose higher energy costs, the costs from a cap-and-trade program would be more difficult to estimate, because the market would determine the price of emission allowances (and thus the overall costs of the program). In other words, cap-and-trade proponents may have an advantage in promoting their approach, because they can cite studies that estimate relatively low program costs.[57]

Administrative Issues

A well-developed administrative structure for collecting taxes already exists in the United States. Moreover, fuel sales are well-documented and are currently taxed (for various reasons) to some degree.[58] Thus, a carbon tax on fossil fuels—if applied upstream in the economy (discussed below)—could utilize this existing framework. However, a carbon tax that included other GHG emissions or their inputs would not have the advantage of this existing administrative structure. Point of taxation issues are discussed later in this chapter.

Some argue that an emissions trading system would require the establishment of new institutions to administer the program.

> [T]he upfront costs of creating institutions to administer trading are significant and likely to produce entrenched bureaucracies that clamor for ever-tighter controls on carbon emissions.[59]

Others maintain that these concerns may be overstated. "Experience with existing trading programs, such as the U.S. SO_2 trading program, has shown that these institutions can arise quickly and for the most part inexpensively."[60] However, comparing the SO_2 trading program with a potential GHG emission trading program highlights appreciable differences. First, the scale of the SO_2 trading program is considerably smaller than the proposed cap-and-trade initiatives. Second, the SO_2 emission sources are more concentrated than GHG-emitting sources, which cut across multiple economic sectors.

As with other comparisons, the relative advantage of a carbon tax would depend on the designs of the instrument alternatives under scrutiny. For example, a cap-and-trade system with a more downstream structure may present more of a challenge, because more sources would be subject to the cap. In addition, if a cap-and-trade program were to provide emission allowances to covered sources at no cost, policymakers would need to devise a system for determining allotment. Emission allowance auctions would alleviate this responsibility, but would require an additional administrative responsibility, and some have raised concerns that auctions may impose a risk of market manipulation.

Another potential administrative advantage of a carbon tax approach is that it carries less risk of being hindered by litigation. Compared to a carbon tax, cap-and-trade legislation may delegate more authority to EPA or another government agency to implement certain components—e.g., specific allocation decisions, offset protocols, etc.—of the program.[61] Such decisions may be vulnerable to legal challenges from stakeholders.

Policy Modification

Some have argued that one of the advantages of a carbon tax is the relative ease—compared to a cap-and-trade program—in which the program's stringency could be modified.[62] In contrast, they assert that policymakers would face difficulties if they sought to adjust an emissions cap after the program's initiation.[63] The rationale for this assertion is that covered sources that made or purchased emission allowances beyond those needed in a given year would lose some of the value of these allowances if Congress raised (i.e., loosened) the cap at a later time. Similarly, a covered source may make capital investments based on the assumption of a stringent cap. If policymakers subsequently loosened the cap, these covered sources would take longer to recoup their investments.[64] However, this concern could also apply to a carbon tax. For example, energy producers and consumers may make investments based on an expected carbon tax. If the tax is subsequently altered, the value of such investments may change.

POTENTIAL DISADVANTAGES

A GHG emission reduction program that employs a carbon tax may face several disadvantages, compared to a cap-and-trade program. As with the potential advantages, the size and/or relevancy of the disadvantages are dependent on the design of the programs being compared. The potential disadvantages are discussed below.

Uncertain Emissions

The primary disadvantage of using a carbon tax to control GHG emissions is that the level of emissions would be uncertain.[65] For some this concern may present a non-starter, precluding a carbon tax as an option to control GHG emissions. As discussed above, some argue that the potential for irreversible climate change impacts necessitates the emissions certainty that is only available with a quantity-based instrument (e.g., cap-and-trade).

However, uncertain emissions do not necessarily equate with no emission control. Multiple models have estimated the carbon tax rate that would be required to achieve certain emission targets, targets that are comparable with the emission reduction goals of recent cap-and-trade proposals.[66] Regardless, these models can only provide estimates based on the best information available about the marginal costs of abatement (**Figure 1**).

Although uncertain emissions are inherent with a carbon tax approach to emission control, there are policy options available to Congress to enhance the emission control certainty of a carbon tax. In theory, policymakers could devise a carbon tax program that yields only short-term emission fluctuations, as it progresses towards its long-term emission reduction objective. To achieve this goal, Congress would need to enact a mandatory GHG emission reporting regime to act in parallel with a carbon tax. Data from such a regime could be used to track the impact and performance of a carbon tax. If policymakers determine emission reduction is not occurring at a desired pace, the tax rate could be amended.

As discussed above, carbon tax proponents would argue that short-term emission variations would not undermine efforts to control climate change. Indeed, they would assert that short-term emission fluctuations are greatly preferred to the price volatility that a cap-and-trade program would potentially impose. This argument is supported by the notion that CO_2 is a stock pollutant, which generates damages through its overall concentration in the atmosphere, not its annual flow.

Critics of such an approach may question whether modifying the carbon tax rate on a periodic basis would be a politically palatable option. For example, it may be difficult for policymakers to increase the tax rate, especially during downturns in the economy. To address these concerns, some have suggested that Congress authorize an independent board or agency with the ability to modify the tax rate, in order to meet pre-determined emission reduction objectives.[67] Although this approach would improve emission certainty, long-term price certainty would be sacrificed to some degree, depending on the authority of the delegated entity to adjust the tax rate.

Political Feasibility

A common argument against a carbon tax is that it would present more political challenges to enact than an emissions cap, and thus policymakers should focus on the latter instrument. The Council on Foreign Relations recommended a cap-and-trade system over a carbon tax, stating:

> The Task Force finds that, assessed from a domestic economic perspective, the relative advantages of cap-and-trade and carbon tax approaches are easily overstated. Either, designed with a mix of efficiency and fairness in mind, would be an acceptable approach to reducing emissions. The Task Force finds, however, that the *political momentum* behind capand-trade makes its near term adoption much more likely.... [emphasis added][68]

Although a carbon tax would likely face more political obstacles than a cap-and-trade program, some of these obstacles may be based on misunderstandings regarding the differences between the two approaches or on assumptions that the tax would be set too low to be effective. Carbon tax proponents may be able to address these issues to some degree, but the recent political momentum for a cap-and-trade program may trump these efforts. This section examines some of the political concerns associated with a carbon tax.

What's in a Name?

Enacting a "carbon tax" to control GHG emissions would pose a political challenge for policymakers simply due to the word "tax." In contrast, the term "cap-and-trade" does not carry the same stigma. To some degree, the terminology advantage of a cap-and-trade program is arguably unfair, because it would act very much like a carbon tax. By placing a price on carbon, both a cap-and-trade and a carbon tax would likely impose costs on the economy. Indeed, some Members have argued that a cap-and-trade approach should be labeled as a "cap and tax," because it would act effectively as a tax.[69]

Proponents of enacting a carbon tax understand the political ramifications of their instrument's name. To overcome this hurdle, some have described carbon tax programs as "user fees" or "tariffs"or if they are highlighting revenue distribution opportunities—carbon tax "swaps"[70] or "shifts." There may be procedural consequences of a name change—see the text box "The Role of Committee Jurisdiction," below. Proponents of a carbon tax could potentially make inroads in the debate by framing the debate within the larger context of tax policy, particularly tax reform.

THE ROLE OF COMMITTEE JURISDICTION

The instrument's name and function would play an important role in terms of committee jurisdiction. A carbon tax proposal would likely be referred to the House Committee on Ways and Means and/or the Senate Committee on Finance; a cap-and-trade proposal would likely be referred to a committee that covers environmental policy (e.g., the Senate Committee on Environment and Public Works, House Committee on Energy and Commerce). In general, environmental policy committees may have different perspectives, expertise, or priorities than tax committees. These differences could influence the design of a GHG control instrument that originates in a particular committee, as well as subsequent program oversight and adjustments: e.g., the use of revenue from a carbon tax or a cap-and-trade system. The committee jurisdiction factor may play some role in the conceptualization of, and the debate over GHG emission control.

Some may contend that if the policy were called a "user fee" or "user charge" then the primary committee of jurisdiction would not necessarily be a tax policy committee. Standing committees of the House, other than the Appropriations and Budget Committees, may report legislation creating or modifying user fees. However, this procedural point would likely be the subject of intense debate: A ruling on whether a policy instrument is a user fee or a tax measure may depend on the nature of the charge rather than its label. For more information on this issue see U.S. Office of Management and Budget, Analytical Perspectives, Budget of the United States Government, Fiscal Year 2009 (2008); and Congressional Budget Office, The Growth of Federal User Charges (1995).

Some Members have argued that a cap-and-trade proposal would ultimately act as a tax on GHG emissions. They question whether the potential of a cap-and-trade program to raise revenue would have implications for committee jurisdiction. See Letter from Representatives Sensenbrenner and Cantor to Representatives Conyers and Nadler (February 11, 2009), at http://republicans.globalwarming.house.gov/index.shtml.

Support from industry?

Certainly the term "industry" covers a wide and diverse array of economic actors. Their motivations for support of, or opposition to, a carbon tax may vary as well. Support may depend on the anticipated design of the alternative—i.e., a cap-and-trade system. Cap-and-trade proposals have evolved in recent years. Industry stakeholders may have supported a cap-and-trade approach because, if enacted, they expected to receive some or perhaps all of the allowances at no cost.[71] This may not be an unrealistic expectation, considering the emission allocation strategies in the sulfur dioxide (SO2) emissions cap-and-trade program and the European Union's Emissions Trading Scheme (EU ETS),[72] both of which provide almost all of their allowances to covered sources at no cost. The no-cost allowance allocation to EU emissions sources generated windfall profits for some industries.[73]

However, the more recent cap-and-trade proposals from the 110th Congress would have provided considerably fewer allowances to covered sources at no cost.[74] Industry stakeholders have voiced concern over the apparent evolution of allocation strategy.[75] If industry expects to receive fewer allowances at no cost than previously thought, their support of a cap-and-trade system may erode, which could increase interest in a carbon tax.

Some argue that if industry is acting in its own self-interest, taxes would be the least-preferred option, even behind auctioned allowances in a cap-and-trade system.[76] The explanation for this preference is that auctions may allow room for industry to manipulate the price, driving down the overall cost of compliance (as opposed to a tax, which would be fixed). But some industries would prefer the cost to be fixed and known in advance (as a carbon tax would do), so they can better assess investment options (e.g., equipment upgrades, efficiency improvements, etc.).[77] In addition, some industry stakeholders may lend their support to a tax, if the alternative were to resemble recent cap-and-trade proposals (from the 110th Congress), which would have applied directly to upstream entities (i.e., energy producers). This approach would resemble a tax— without a price certainty—for downstream energy consumers (e.g., power plants).[78] Moreover, the upstream entities subject to direct compliance—i.e., required to submit allowances—would have little or no direct emissions. The gap between compliance and emissions could pose challenges in a situation without a known price (i.e., a cap-and-trade program).

Support from environmental groups?

Supporters of stringent GHG emission control may oppose a carbon tax, because they assume that if Congress enacts a carbon tax, Members would set the rate too low to induce emission mitigation: e.g., energy conservation, fuel substitution, or technology development. This may be more of an argument against the tax rate, than the instrument itself.

A similar concern of stringency may exist if policymakers decide to employ a cap-and-trade system. For example, the cap might be set at undesired levels, or the program may allow for the use of questionable offsets, which could pose a risk of undermining the objective of emission reduction. Likewise, the pressure to delay the strengthening of an established cap that is considered ineffective or expensive could be considerable. Regardless, an emissions cap—at least on paper—offers more of a guarantee that a specific emission level would be met. Some environmental groups may assert that this perceived guarantee overshadows other concerns.

Consideration of International Efforts and Cooperation

It is generally recognized that mitigation of human-induced climate change will ultimately require action by all major emitting countries. Therefore, when assessing different mechanisms of GHG emission reduction, Congress may want to consider how an market-based instrument—e.g., a cap or a tax—would 1) complement existing or future international efforts and/or 2) maximize participation among major emitting nations.

For policymakers and negotiators who are attempting to create an international GHG reduction regime, establishing either instrument on an international scale would present substantial and unique challenges. A carbon tax may be at a disadvantage, primarily because the most prominent international activities currently involve cap-and-trade programs. However, some assert that a carbon tax approach may present a stronger opportunity to maximizc participation between the United States and developing economies—e.g., China and India—that emit substantial portions of GHG emissions. Some of the issues are discussed below.

Coordination with existing international efforts

At the core of the international GHG mitigation efforts underway—e.g., the Kyoto Protocol and the EU ETS—is a cap-and-trade program among nations with quantitative commitments. If Congress enacted a cap-and-trade system in the United States, the program could be designed to allow trading between systems. In terms of overall economic efficiency, linked trading programs would be ideal, because covered sources would have access to more emission reduction opportunities. However, linked programs could yield unintended consequences if they differ in scope and/or stringency.[79]

A U.S. carbon tax program could conceivably have a linkage of sorts with international cap-andtrade programs. For example, a carbon tax system could allow for tax credits for emission reductions made outside of the United States.[80] However, this trade would only flow in one direction: Covered sources in the EU could not purchase reductions from covered sources in the United States. Thus, this type of linkage would not equate with a union of cap-and-trade programs. Sources in the United States that were subject to the carbon tax would find this opportunity worthwhile only if the carbon tax rate were higher than the emission allowance price (EU emission allowances or EUAs) or, if eligible, offset project prices (traded as certified emission reductions or CERs) in the EU ETS. However, if policymakers were to provide tax credits for international emission abatement projects, concerns similar to those involving offsets in a cap-and-trade system would be raised. If the illegitimate projects qualify as tax credits, the primary objective of the program—GHG emission control—may be compromised.

Maximizing participation

Some have raised the argument that encouraging maximum participation among nations should be a higher priority than instrument choice and the potential consequences of choosing a cap over a tax (e.g., loss of economic efficiency).[81] Would a carbon tax or a cap-and-trade program be the most effective approach to bring the most nations, or more precisely, emissions, under a mitigation umbrella? Neither strategy would necessarily require equal commitments between nations. For example, the United States and China could each establish unilateral carbon tax programs—as opposed to an international, harmonized system—with different tax rates. Indeed, an international carbon tax system would encounter substantial implementation challenges, particularly regarding issues of national sovereignty.

It would be (and has proven) difficult to achieve wide and diverse participation—namely, agreement from both the United States and developing nations—in a GHG control regime. Some may favor a cap-and-trade approach, because of the buildup of existing programs and the political momentum behind the cap-and-trade approach. However, this reasoning may be questioned on several counts. First, although a substantial percentage of the world's CO_2 emissions are covered under the Kyoto or EU ETS cap, an even larger percentage are not covered.[82] This latter group includes the United States, China, and India, which accounted for approximately 54% of the CO_2

emissions from the top-25 CO_2 emitting nations in 2005.[83] Second, the mere existence of a program (and its political momentum) does not indicate it is the optimal solution. Indeed, the existing international cap-and-trade systems, which are still in their early stages, have encountered substantial problems.[84]

However, a carbon tax approach would encounter obstacles similar to those observed in the capand-trade discussions. For example, it would be difficult to reach agreement on the equitable or fair tax rates for each nation. Not only would this agreement be difficult between developed and developing nations—for multiple reasons that are beyond the scope of this chapter—but it would also pose a challenge between developed nations. For example, a 2000 study found that European nations were (at the time of the study) effectively taxing carbon at a rate of approximately \$27/ton of CO_2.[85] If European nations and the United States sought to establish their own carbon tax rates, pre-existing carbon taxes (and subsidies) may complicate the discussions.

International implementation concerns

Regardless of the chosen instrument, either would impose its own set of logistical challenges. Some assert that an international cap-and-trade system would be more susceptible to corruption.[86] However, it could be argued that monitoring and enforcement of carbon taxes in individual nations might pose similar difficulties. A primary obstacle would be determining a nation's "net carbon tax," because nations or political subdivisions could offset the carbon tax through energy subsidies or loopholes for the entities most impacted by the tax.[87] Moreover, some nations may not have transparent tax systems in place now. Adding a carbon tax to such a framework could be problematic.

IMPLEMENTATION OF A CARBON TAX

If Congress were to use a carbon tax to control GHG emissions, policymakers would face several implementation decisions: (1) the point of taxation—where to impose the tax and what to tax; (2) the level of taxation; and (3) how to apply the tax revenue. These considerations are discussed below.

Point of Taxation

A point of taxation debate is analogous to a debate over the scope or coverage of a potential capand-trade program. Congress would face similar considerations of emissions coverage with either instrument. Describing coverage issues, EPA stated:

> Ideally, all sources, sectors, and emissions would be included for full coverage and maximum environmental effectiveness and economic efficiency. However, measurement capabilities and costs, available control options, administrative burdens, political considerations, and other constraints may limit participation to a subset of emission sources.[88]

A comprehensive discussion of these issues is beyond the scope of this chapter,[89] but the following sections provide options for policymakers. Note that these sections overlap to some degree.

Where to impose a carbon tax?

Regarding where in the economy to impose a carbon tax, the options—at least for fossil fuel combustion-related CO_2 emissions 90—are often couched in terms of their location in the energy production and consumption chain: upstream or downstream (**Figure 7**):

- **Upstream**—An upstream approach would apply a carbon tax to fossil fuels when they enter the U.S. economy, either at the mine, wellhead, or another practical "chokepoint" in the production chain. Imported fuels would be addressed at their point of entry into the United States (**Figure 7**). Under this approach, a carbon tax would be levied before emissions have occurred. The tax would be based upon the carbon content of the fossil fuel produced. An advantage to the upstream approach is that the number of sources directly subject to the tax would be minimized, while covering a broad spectrum of the economy.

- **Downstream—**A downstream approach would apply the tax on the emissions themselves. For example, emission sources, such as power plants or industrial facilities,[91] would pay a fee for each ton of CO2 released into the atmosphere. In some economic sectors, there may be advantages to imposing a carbon tax downstream, even if more entities would be directly subject to the tax. For instance, Congress may consider a downstream approach for the electric utility industry, because power plants already continuously monitor their CO2 emissions to the EPA.[92] There may be other advantages to applying the tax directly to emissions from coal-fired power plants. For example, some argue that this approach would provide a greater stimulus[93] to the development of carbon capture and sequestration (CCS) technology.[94] However, in some sectors—e.g., transportation or residential—a downstream approach may be challenging, and perhaps impractical to implement because of the comparably large number of emission sources.

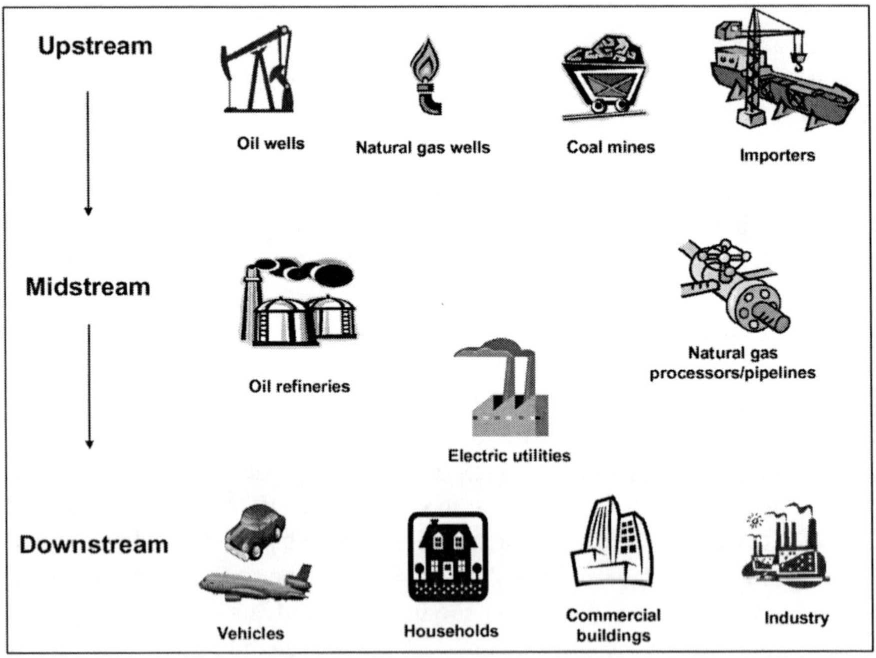

Source: Prepared by CRS.

Note: Electric utilities could be listed as either downstream entities—because they are direct sources of emissions—or midstream, because their emissions are tied to the electricity consumption of their customers, the further downstream consumers.

Figure 7. Illustration of Options for Points of Taxation within the Energy Production-to-Consumption Chain

- **Combined Approach**—Sometimes referred to as a hybrid strategy, this approach would cover emissions from different economic sectors in different manners. For example, transportation fuels could be taxed upstream at the point of extraction (or midstream at the refinery); coal could be taxed at the emission sources: power plants and specific industrial sectors.

Where Congress decided to impose a carbon tax would determine which entities would be required to (1) make tax payments based on emissions or emission inputs, (2) monitor emissions or emission inputs, and (3) maintain records of relevant activities and transactions.

A direct tax on emissions may be easier to understand, but considerably more difficult to implement across all economic sectors, due to the large number of emission sources. A complicating factor is that some fossil fuel uses—e.g., feedstock in asphalt production—do not generate emissions but sequester the carbon. Policymakers may consider mechanisms to exempt the fuels used for such purposes or allowing for tax credits for these activities.

Fossil fuels have varying levels of carbon content, thus a carbon tax would impact fossil fuels differently. For example, coal has almost twice the carbon content per unit of energy as that of natural gas (**Table 1**). Thus, a carbon tax would raise the price of coal more than the price of natural gas.

Table 1. CO_2 Emissions Per Unit of Energy for Fossil Fuels

Fossil Fuel	CO_2 Emissions Per Unit of Energy (million metric tons/quadrillion BTU)
Coal	94
Crude oil	74
Natural gas	53

Source: Prepared by CRS, based on CO_2 emission factors from EIA, at http://www.eia.doe.gov/environment.html.

Note: CO_2 emissions from coal vary by the type of coal being used.

THE 1993 ENERGY OR BTU TAX

A carbon tax is different from an energy consumption tax, like the British thermal unit (Btu) tax that was proposed by President Clinton in 1993. Compared to a carbon tax, a Btu tax would be a less efficient mechanism to reduce CO_2 emissions. As highlighted in, fossil fuels generate different amounts of CO_2 emissions per unit of energy. A Btu tax would treat energy sources equally, thus a Btu tax would serve different objectives than a carbon tax.

The 1993 Btu tax proposal called for a levy of 25.7 cents per million Btu, with a surcharge of 34 cents/million BTU on petroleum. The goals of the 1993 Btu tax proposal were to promote energy conservation and raise revenue. At the time, the proposed tax would have generated a new revenue stream of about $30 billion per year. The proposal was met with strong opposition and was not enacted; Congress ultimately enacted an (approximately 5-cent) increase in the motor fuels taxes.

CO_2 emissions or All GHG emissions?

Another point-of-taxation issue is whether a carbon tax should cover CO_2 emissions or a more comprehensive array of GHG emissions. Mitigation opportunities from non-CO_2 gases are often less expensive than other reduction options. Economic studies indicate that one advantage of including more or all GHG emissions, as opposed to only CO_2 emissions, is that the overall costs of the GHG reductions would be decreased.[95] A recent analysis of a carbon tax proposal—H.R. 3416 (Larson)—from the 110th Congress found that by adding other GHG emissions to the tax base, the overall costs of achieving the same amount of emission reduction would fall by 20%.[96] However, this finding was based on the inclusion of GHG emissions from all sources. As discussed below, the majority of the non-CO_2 GHG emissions come from sources that may present tax implementation challenges.

Which emissions sources to control?

GHG emissions are spread throughout the economy and are generated by millions of discrete sources: e.g., smokestacks, exhaust pipes, households, and livestock. Although CO_2 is the primary GHG, some GHG emissions sources predominantly emit non-CO_2 GHGs. When determining which sources and gases to control through a tax, policymakers would need to balance the benefits of inclusion with administrative costs.[97] To achieve a specific emissions target, a tax applied on only a subset of economic sectors—thus covering only a percentage of U.S. GHG emissions—would need to have a higher rate than a tax applied to all GHG emissions. Numerous rationales have been put forward to justify excluding one group or sector from a reduction requirement, or to provide some other special consideration.

Table 2 lists the top emission sources of six GHGs in the United States. These sources combine to account for approximately 95% of the total (based on 2006 data).[98] **Table 2** also provides potential points of taxation that would cover the emissions from these sources.

CO_2 emissions from fossil fuel combustion yielded approximately 80% of U.S. GHG emissions in 2006.[99] Policymakers could address CO_2 emissions from fossil fuel combustion and non- energy uses—in aggregate 82% of U.S. GHG emissions—by levying an upstream carbon tax on fewer than 2,500 upstream or midstream entities (**Table 2**). A portion of these entities are already subject to environmental taxes that support other objectives.[100]

A downstream approach may be feasible for electric generating facilities powered by fossil fuels, which number approximately 3,300,[101] but impractical for other fossil fuel combustion sources— automobiles, offices, homes—which number in the millions.

One possible alternative to a complete upstream/midstream strategy would be to address coal, the most carbon-intensive fossil fuel, at its combustion location. In 2006, coal-fired power plants accounted for 92% of U.S. coal combustion.[102] As shown in **Table 2**, there are about 641 coal- fired power plants in the United States. Some argue that there would be advantages to applying a point of taxation at electric generating facilities. For example, these facilities may have a stronger incentive and/or capability to seek tax credit opportunities, such as biological sequestration projects (if allowed within the tax system). Moreover, some contend that the development of CCS would be enhanced if coal-fired facilities were directly subject to the tax. The rationale for this argument is that under an upstream approach (that indirectly taxes coal-fired emissions) the price signal may be weaker than if the tax were applied directly to emissions at coal-fired facilities. However, other sources of coal combustion would need to be covered to avoid leakage situations—i.e., increased coal (and thus emissions) use in the uncovered sectors. It may be burdensome to address the remaining sources of coal combustion at the point of emissions, particularly at residences and commercial entities.[103]

Beyond fossil fuel combustion, policymakers could cover more GHG emissions by expanding the coverage to some or all of the remaining sources listed in **Table 2**. Individually, many of these sources account for relatively small proportions of U.S. GHG emissions, but if some combination of these sources were included in the carbon tax base, the increase in emissions coverage could be appreciable.

The remaining source categories vary by (1) the number of entities that would be subject to a carbon tax, and (2) the circumstances of emission release. In general, a large number of entities subject to a carbon tax would present higher administrative costs per unit of emissions. Regardless of entity number, some emissions sources would be more difficult to cover because of logistical challenges, such as emission measurement and/or monitoring.

Table 2. Selected Sources of U.S. GHG Emissions and Potential Applications of a Carbon Tax

GHG Emission Source	Percentage of U.S. GHG Emissions (2006 data)	Potential Carbon Tax Applications	
		Entity	Number
CO_2 from fossil fuel combustion:	79.9	Coal mines[a]	1,374
• electricity generation		or	641
• transportation		Coal-fired power plants[b]	
• industrial		Power plants using [c]	26
• commercial/residential		imported coal [d]	
		Petroleum refineries [e]	150
		Petroleum importers [f]	220
		Natural gas processors[g]	530
		Natural gas importers [h]	45
N_2O from agricultural soils	3.8	Farms[i]	>2 million
CO_2 from non-energy use of fuels	2.0	Covered by the tax applied to fuels (above)[j]	
CH_4 from livestock (enteric fermentation)	1.8	Cattle operations [k]	967,440
CH_4 from landfills	1.8	Landfills [l]	1,800
HFCs from the substitution of ozone depleting substances	1.6	HFC manufacturers [m]	5
CH_4 from natural gas systems	1.5	Natural gas processors	530
CH_4 from coal mines	0.8	Coal mines [n]	1,374
CO_2 from iron/steel production	0.7	Raw steel production facilities; [o]	116
		Integrated steel mills	18
CO_2 from cement manufacturing	0.6	Cement plants[p]	118
CH_4 from manure management	0.6	Cattle operations;	967,440
		Swine operations[q]	65,640
Percentage of Total GHG Emissions	95.1		

Source: Prepared by CRS; GHG emission data from EPA, EPA, *Inventory of U.S. Greenhouse Gas Emissions and Sinks: 1990-2006* (April 2008); data for number of entities from multiple sources, cited in footnotes.

[a] This figure accounts for mines in operation in 2007. EIA, *Coal Production and Number of Mines by State and Mine Type*, (2008).

[b] In 2006, there were 641 plants with at least 1 coal-fired generator unit (EIA, 860 Database).

[c] Number of plants comes from EIA database, Monthly Nonutility Fuel Receipts and Fuel Quality Data (Database 423). CRS was unable to determine the number of companies that act as coal importers, analogous to petroleum importers.

[d] In 2006, the United States imported approximately 36 millions short tons of coal (EIA, Quarterly Coal Report (2008), table 4)—3.5% of the amount to coal consumed domestically in that year (EIA, Annual Coal Report 2006 (2007), table 26). Coal imports have increased by more than 200% since 2002.

[e] This figure represents the number of "operable" refineries. *EIA, Refinery Capacity Report* (2008).

[f] EIA, Company Level Imports (as of November 2008). Note that some of these companies may import only crude oil, whose emissions would be covered by the tax at the domestic refineries. Thus, this figure represents an upper bound of petroleum product importers potentially subject to a carbon tax.

[g] EIA, Natural Gas Processing: The Crucial Link Between Natural Gas Production and Its Transportation to Market (2006).

[h] This includes pipelines and liquefied natural gas facilities. EIA, *About U.S. Natural Gas Pipelines* (as of September 2008).

[i] U.S. Department of *Agriculture, Farms, Land in Farms, and Livestock Operations: 2007 Summary* (2008). The resource defines a farm as "any place from which $1,000 or more of agricultural products were produced and sold, or normally would have been sold, during the year."

[j] Fossil fuels are used for a wide range of non-energy purposes. EPA estimates that of the total carbon consumed for non-energy purposes, approximately 62% is stored in products, and not released to the atmosphere (EPA, Inventory of U.S. Greenhouse Gas Emissions and Sinks: 1990-2006 (April 2008), tables 3-14 and 3-15). The 2% value in Table 4 represents the emissions. In an upstream carbon tax system, fuels would be taxed before they are used. Congress may want to consider providing tax credits for the amount of carbon stored in products.

[k] U.S. Department of Agriculture, Farms, Land in Farms, and Livestock Operations: 2007 Summary (2008).

[l] EPA, Inventory of U.S. Greenhouse Gas Emissions and Sinks: 1990-2006 (April 2008), citing BioCycle, 15 th Annual BioCycle Nationwide Survey: The State of Garbage in America (2006).

[m] Intergovernmental Panel on Climate Change, Safeguarding the Ozone Layer and the Global Climate System, Issues related to Hydrofluorocarbons and Perfluorocarbons (2005), Figure 11.1.

[n] Methane from underground mines, which accounts for about 61% of coal mine methane, is removed through ventilation systems for safety reasons. These emissions would be easier to monitor under a carbon tax than aboveground coal mine methane emissions.

[o] Data from U.S. Geological Survey, Mineral Commodity Summary, Iron and Steel Production (2008), at http://mineralspubs/commodity/ iron_&_steel/.

[p] Cement manufacturing information from Portland Cement Association, at http://www.cement.org/basics/cementindustry.asp.

[q] U.S. Department of Agriculture, *Farms, Land in Farms, and Livestock Operations: 2007 Summary* (2008). Other animals—chickens, horses, and sheep—contribute approximately 10% of the total emissions from manure (EPA, Inventory of U.S. Greenhouse Gas Emissions and Sinks: 1990-2006 (April 2008), table 6-6).

Although GHG emissions from the agriculture sector—farming and livestock—account for approximately 6% of total GHG emissions in the United States, agricultural emissions would be comparatively difficult to cover under a carbon tax. As indicated in **Table 2**, these emissions are generated by a large number of entities, and there are no upstream measurement points analogous to those that could cover emissions from the transportation sector. However, nitrous oxide (N2O) emissions from the use of artificial fertilizers—as opposed to natural fertilizers like manure— might be an emission source that could be addressed through an upstream point of taxation. EPA estimates that approximately 25% of the N2O emissions from agricultural soils are related to artificial fertilizer use.[104] This represented approximately 0.8% of U.S. GHG emissions in 2006, so it would be included in **Table 2** if counted separately. A tax on fertilizers may lead to less fertilizer use, but some argue that it could stimulate other activities—e.g., increased use of natural fertilizers—that would release N2O emissions (a form of emissions leakage).[105] This issue is beyond the scope of this chapter.

In many cases, the emissions from agricultural activities may be difficult to monitor and/or measure. For practical concerns and other reasons—e.g., political considerations—the agriculture sector has thus far been excluded as a capped source in the cap-and-trade proposals that have specifically designated covered sources.[106] If not subject to a carbon tax, agriculture emissions could be addressed with other policy tools. For example, tax credits could be provided for qualifying reduction, abatement, or sequestration projects.[107]

The other emission sources listed in **Table 2** may be more amenable to a carbon tax, at least in terms of the number of entities and emission circumstances. Compared with agricultural sources, the other emission source categories—excluding HFCs (discussed below)—comprise a relatively small number of entities, at which emissions are easier to measure and monitor. In fact, certain sources—e.g., some landfills and underground coal mines—currently control their (methane) emissions for air pollution and safety reasons.

Methane from landfills—1.8% of U.S. emissions—could be covered by including an additional 1,800 landfills in the tax base. To some degree, large municipal landfills are already required per federal air regulations to collect and combust methane emissions. Other landfills are following suit under state laws or voluntary programs.[108]

Hydrofluorocarbons (HFCs) are used as substitutes for several classes of ozone depleting substances (ODSs—hereafter, HFC-ODS).[109] HFC-ODS account for 1.6% of total U.S. GHG emissions (**Table 2**). HFC-ODS emissions occur under different circumstances than the other, "smaller" sources discussed above—landfills, coal mines, etc. Most of the HFC emissions (86% in 2006) come from air conditioning and refrigeration units—located in vehicles, buildings, and homes—during production, operation, and disposal.[110] Unlike the other "small" source categories, which emit GHGs onsite, HFC-ODS emissions enter the atmosphere from millions of locations, far away from and long after HFC production. Because of the different emission circumstances of HFC-ODS, some have proposed combining a tax at HFC production—five facilities in the United States—with a tax credit for HFCs properly recycled or disposed.[111] For example, automobile air conditioners could be sent to designated sites in order to recycle or dispose of the HFCs. Norway has an HFC rebate program in place that may be instructive for U.S. policymakers.[112] As with other carbon-intensive materials, Congress may consider applying a tax to imported HFCs as well.

By applying the tax to iron/steel production and cement manufacturing facilities—approximately 250 entities—an additional 1.3% of U.S. emissions could be covered.[113] A further 1.5% could be addressed by including methane from natural gas systems. Roughly 25% of this amount comes from field production, which may be impractical to monitor and measure accurately.[114] The remainder involves emissions from normal operations, routine maintenance, or "system upsets." Fugitive emissions (or accidental releases) are the primary source of emissions from these activities.[115] A tax on these emissions may impose administrative costs in terms setting up a process to measure and monitor these emissions.

Methane emissions from underground coal mines—0.8% of total GHG emissions—may be easier to monitor under a tax than at aboveground mines. At all underground mines, methane is ventilated for safety reasons; methane from aboveground operations is released directly to the atmosphere. The 0.8% figure includes methane only from underground coal mines, which account for about 61% of methane from coal mines.[116]

Level of Taxation

If Congress were to use a carbon tax to control GHG emissions, setting the level of the carbon tax would be a subject of intense debate. Regardless of the chosen tax schedule, some parties would argue that the selected rate was either too low or too high. Although the decision would likely involve political calculations, there are economic principles that could be used to inform the debate. This section identifies and discusses two approaches that policymakers could take to set the level of a carbon tax. The first approach is based on the estimated benefits associated with avoiding climate change impacts. The second approach is based on a GHG emissions target and the tax rate needed to achieve that target. The first approach is arguably stronger from a theoretical standpoint, but the second approach is arguably (considering the uncertainty associated with the first approach) a more practical option.

To provide some context for potential carbon tax rates, **Table A-1** of this chapter (located in the **Appendix)** includes estimates of price increases to fossil fuels and motor gasoline based on different carbon tax rates.

Tax based on estimates of costs and benefits[117]

As noted previously, economic principles involving pollution taxes prescribe that the optimal tax should be at the intersection of the marginal costs of emission abatement with the marginal benefits of abatement. In terms of climate change, the intersection of costs and benefits is often described as the Social Cost of Carbon (SCC). The Intergovernmental Panel on Climate Change (IPCC) defines the SCC as the following:

> The discounted monetized sum (e.g. expressed as a price of carbon in $/t$CO_2$) of the annual net losses from impacts triggered by an additional ton of carbon emitted today. According to usage in economic theory, the social cost of carbon establishes an economically optimal price of carbon at which the associated marginal costs of mitigation would equal the marginal benefits of mitigation.[118]

For multiple reasons, applying this economic principle to set a carbon tax currently presents monumental, if not insurmountable, challenges. These are highlighted below.

Social cost of carbon estimates

Estimates of the SCC are fraught with uncertainty, thus the wide range of estimates produced should not come as a surprise. According to a 2008 meta-analysis, there are over 200 SCC estimates from 47 studies. The estimates range across several orders of magnitude, from zero to over $500 per ton of CO_2-equivalent (tCO2-e) emissions.[119] The uncertainty and controversy of choosing a value within this range would be formidable.

The definition of SCC indicates the complexity involved in its estimation. The SCC can be defined as a monetary measure of the marginal damages (or costs)—on a global scale—of an incremental unit (e.g., one metric ton) of anthropogenic GHGs emitted to the earth's atmosphere at some point in time (e.g., now).

Estimating the SCC requires analysts to place values on goods and services that may be difficult (or controversial) to precisely measure: e.g., human health/life, water supplies, agricultural production, recreational activities. The value of these goods/services will vary from one location to the next, complicating the estimate. Moreover, the climate change processes that may impact these goods and services are not completely understood.[120]

The element of time particularly complicates an SCC estimation and application in several ways. An SCC estimate signals—per economic theory—what global society should be willing to pay *now* to avoid *future* damages due to additional emissions generated *today*. As discussed above, CO2 and other GHGs remain in the atmosphere for decades to centuries, thus current emissions could generate damages for years to come. To take time into account—i.e., determine the cost of future damages in today's dollars (referred to as net present value)—economists employ a discount rate.[121] Discount rates are typically at the crux of cost-benefit analyses and their assignment often generates controversy. As with other public policy decisions that employ cost-benefit analysis, the chosen discount rate is a critical factor when estimating the SCC. In terms of uncertainty in evaluating damages from climate change, the Intergovernmental Panel on Climate Change (IPCC) estimates that the discount rate is the second most important factor (climate sensitivity assumptions are number one).[122]

An additional factor included in SCC estimates is the assumption of the current global emission trajectory. As stated in the Stern Review, "an SCC curve cannot be drawn, nor an SCC calculated, without specific assumptions on future paths."[123] A business-as-usual (BAU) path assumption will produce a higher SCC estimate than an assumed path of emission stabilization.[124] A higher concentration pathway yields a higher SCC estimate, because if an extra ton of GHG emissions is added to an atmosphere with a high GHG concentration, the extra ton is projected to yield more damages than if it were added to an atmosphere of lower GHG concentration.[125]

Applying social cost of carbon estimates

Using SCC estimates as a basis for a carbon tax would be consistent with economic principles, but extremely difficult in practice. The primary reason is that the estimates vary widely, based on the underlying assumptions used to generate the estimates. Selecting one estimate over another may be difficult to justify. For these reasons, economic reports often discuss the SCC in terms of a "starting place"[126] or a "good guide"[127] for setting a carbon tax level.

Moreover, the estimates are snapshots for a given point in time. As conditions change, the SCC estimates change. Thus, using them as a basis for a carbon tax rate may be problematic, because policymakers would need to alter the tax schedule based on the most recent estimates. However, this concern would exist for a carbon tax system based on any economic estimates, such as marginal cost curves (as discussed below).

Regardless, the tax rates from several carbon tax proposals in economic literature[128] and one proposal from the 110th Congress—H.R. 3416 (Larson)—are close in value to the mean (average) of SCC estimates identified in a recent meta-analysis.[129] In 2007, the IPPC found that *peer- reviewed* estimates[130] of the SCC in 2005 dollars have an average of \$12/tCO_2-e.[131] However, prominent economists have generated estimates above and below this level.[132]

Tax based on meeting an emissions target

An alternative to the above approach—basing a carbon tax on expected damages (measured in net present value)—would be to determine the tax rate needed to stimulate a specific emissions reduction goal. For example, Congress would set an emissions target—e.g., 66% below 2005 levels by 2050[133]—and estimate the tax schedule—e.g., \$X/tCO_2-e in 2010, increasing Y% per year—that would encourage the reductions needed to meet the target. Using this approach, policymakers would not need an estimate of the marginal benefits of abatement in the tax rate calculation. However, the chosen emission reduction target may involve some debate over the level of climate change risks that could be tolerated by society.[134]

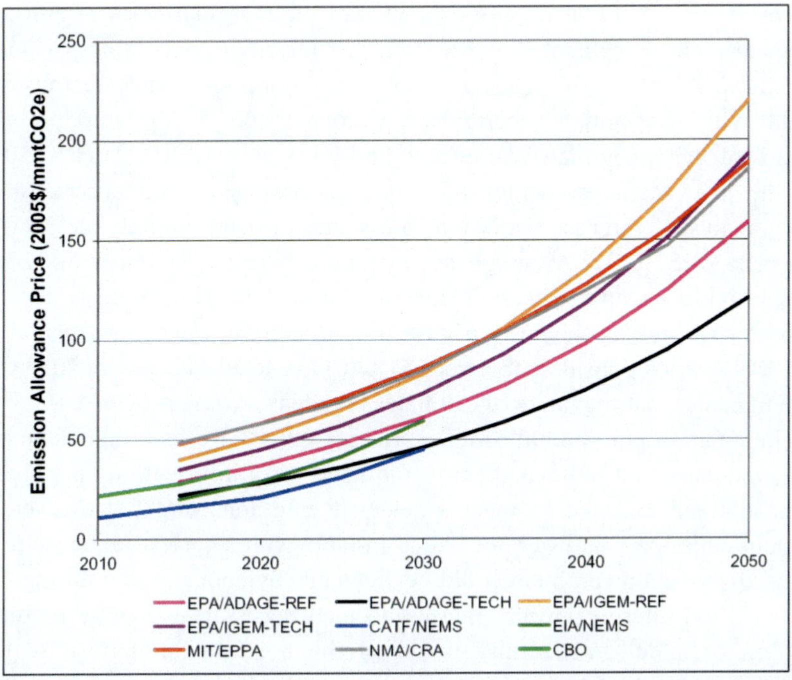

Sources: (1) EPA/ADAGE and EPA/IGEM: "Data Annex" available on the EPA website athttp://www.epa.gov/climatechange/economics/economicanalyses.html;

(2) MIT/EPPA: Sergey Paltsev, et al., "Appendix D" of Paltsev et al., Assessment of U.S. Cap-and-Trade Proposals, MIT Joint Program on the Science and Policy of Global Change (2007);

(3) EIA/NEMS: EIA, Energy Market and Economic Impacts of S. 2191, the Lieberman-Warner Climate Security Act of 2007 (April 2008);

(4) CATF/NEMS: Jonathan Banks, Clean Air Task Force, The Lieberman-Warner Climate Security Act—S. 2191: A Summary of Modeling Results from the National Energy Modeling System (February 2008);

(5) NMA/CRA: CRA International, Economic Analysis of the Lieberman-Warner Climate Security Act of 2007 Using CRA's MRN-NEEM Model (April 8, 2008);

(6) CBO: Congressional Budget Office, Cost Estimate: S. 2191: America's Climate Security Act of 2007 (April 10, 2008).

Notes: This figure was adapted from CRS Report RL34489, *Climate Change: Costs and Benefits of S. 2191/S. 3036*, by Larry Parker and Brent D. Yacobucci. Estimates extrapolated by CRS from available data where necessary. Estimates converted to 2005$ using GDP implicit price deflator. The original figure included two additional estimates that are not included above, because these estimates were subsequently found to contain errors in the underlying methodology.

Figure 8. Emission Allowance Price Estimates under S. 2191

This approach might require estimates of the marginal costs of emission abatement. These estimates, while not as uncertain according to many economists, are similarly based on multiple assumptions, and are thus also fraught with substantial uncertainty. A comparison of the multiple cost analyses of the Lieberman-Warner cap-and-trade proposal (S. 2191/S. 3036) from the 110th Congress demonstrates the uncertainty in estimating costs.[135] These analyses forecast emission allowance prices within the narrow parameters of S. 2191. Although not the same measure, emission allowance prices provide an approximate value of marginal abatement costs for that proposed strategy. As shown in **Figure 8**, emission allowance price estimates for S. 2191 varied considerably, from \$1 6/tCO_2-e to \$49/tCO_2-e in 2015. The figure indicates that the range of estimates broadens as time progresses.

The range of emission allowance prices—a proxy for marginal abatement costs—indicates that policymakers would have weak information on which to set the optimal tax rate to meet a selected emissions target. However, an uncertain emissions pathway goes hand-in-hand with a carbon tax system. As discussed above, this problem could be alleviated by annually monitoring U.S. emissions and altering the tax rate as deemed necessary—thereby requiring Congress or a delegated entity to adjust the tax rate through subsequent legislation or regulation.

Tax Revenue Distribution

Carbon tax revenues would be substantial, and policymakers could apply them to further multiple objectives (not listed in any particular order): allow for reductions in other taxes; alleviate the burden of increased energy prices (brought on by the carbon tax); stimulate development of low- carbon technologies; and support other climate- or non-climate-related purposes. The redistribution of carbon tax revenues is often described as "revenue recycling."[136] After providing an estimate of potential tax revenues for context, this section discusses some of the revenue recycling options and their potential impacts.

Estimates of tax revenues

Table 3 provides estimates of tax revenue estimates from recent legislative proposals from the 110th Congress: H.R. 2069 (Stark) and H.R. 3416 (Larson). Both bills would have imposed a carbon tax on coal, petroleum, and natural gas, based on the CO_2 emissions[137] the fossil fuels would generate upon combustion. As indicated in **Table 3**, the bills were estimated to generate between $69 billion and $126 billion in 2015. To put these estimated tax revenues in context, consider the federal net tax revenue from the three largest revenue sources for FY2005 (in 2005 dollars): [138]

- individual income tax: $1,107 billion;
- employment taxes: $771 billion;
- corporate income taxes: $307 billion.

Uses of tax revenues

Revenues from the carbon tax could be used to promote a wide range of policy objectives, climate-related or otherwise. This section identifies different applications of the carbon tax revenues. A subsequent section discusses the policy considerations for Congress surrounding the different applications of carbon tax revenue.

Offset reductions in other taxes

Carbon tax revenues could be used to replace taxes that apply to desirable activities (as noted earlier). Economic theory generally supports a tax policy that would reduce taxes on favored activities (e.g., increased employment or investment) and increase taxes on less desirable behavior (e.g., increased pollution). The rationale for this policy is that taxes on desirable activities create market distortions, discouraging increased levels of desirable actions. For example, economists argue that income taxes provide a number of distortions to the economy. "By taxing away some of the returns on working and saving, income taxes deter some people from joining the labor force and encourage others to consume too much of their income [instead of saving] ."[139]

Table 3. Estimates of Potential Tax Revenues from Carbon Tax Proposals from the 110th Congress (in 2005 dollars)

	Proposed Tax Rates		**Annual Tax Revenue Estimates**	
Carbon Tax Proposals	**in 2015**	**in 2030**	**in 2015**	**in 2030**
H.R. 2069 (Stark)	\$10/tCO2-e	\$43/tCO2-e	\$69 billion	\$263
H.R. 2069 (Stark)	\$20/tCO2-e	\$83/tCO2-e	\$126 billion	\$361

Source: Prepared by CRS with estimates from Gilbert Metcalf et al., *Analysis of U.S. Greenhouse Gas Tax Proposals* (2008), MIT Joint Program on the Science and Policy of Climate Change, Report No. 160.

Note: The MIT model assumed that H.R. 2069 and H.R. 3416 would start in 2012, although both bills (as written) would have begun in 2008.

Using carbon tax revenues in this manner has been described as yielding a double-dividend: (1) reduced GHG emissions and (2) reduced market distortions from the taxes on desirable behavior. In the early 1990s, some economists suggested that the double-dividend effect would be strong enough to keep the overall costs of carbon reductions to society relatively small or even negative.[140] However, more recent economic studies indicate that the additional tax-imposed costs[141] would probably (at least in the United States)[142] exceed the economic efficiency benefits (unrelated to climate change benefits) from reducing the distortionary taxes.[143]

Direct or indirect payments to households

To alleviate the carbon tax-imposed costs to households, particularly low-income households, policymakers could use the tax revenues to provide financial assistance. There are multiple mechanisms through which households could receive assistance, including lump-sum payments, payroll or income tax deductions, tax credits or rebates, or increases in existing assistance programs.

Provide assistance to carbon-intensive industries

Stakeholders in carbon-intensive industries—e.g., fossil fuel-fired electric utilities, steel producers, paper manufacturing, etc.—argue that if Congress enacts a GHG emission control program, such as a carbon tax, their industries would be adversely impacted. To alleviate potential impacts, policymakers may consider returning—directly or indirectly—a portion of the carbon

tax revenues to affected industries. This would be analogous to providing a percentage of emission allowances (or auction revenues) at no cost to covered sources in a cap-and-trade program.[144]

Fund specific objectives

Congress may also consider using the tax revenues to provide funding to support a range of objectives related to climate change, including:

- Technology development and deployment: Promotion of emission mitigation technology is widely recognized as a vital step towards making substantial GHG emission reductions.[145]
- Energy efficiency: Improvements in energy efficiency could make considerable contributions in achieving GHG emission reductions. Although a carbon tax should stimulate energy efficiency by placing a price on carbon, Congress may consider using the tax revenues to

provide additional incentives, particularly to encourage households to adopt conservation and other measures.

- Biological sequestration: Trees, plants, and soils sequester carbon, removing it from the earth's atmosphere. Tax revenues could be used to promote carbon sequestration efforts: e.g., forestry or agricultural activities.
- Adaptation efforts: Regardless of emission reduction efforts taken today, climatic changes are expected, due to the ongoing accumulation of GHGs in the atmosphere. Therefore, some contend that investment (e.g., tax revenues) should focus on preparing communities—both domestically and internationally—to adapt to the effects of a changing climate.
- Deficit reduction: A portion of the revenues could be set aside to offset projected federal tax revenue shortfalls.[146] In addition, Congress may consider using carbon tax revenues to offset shortfalls beyond those that are related to the carbon tax.

Policy considerations of different revenue applications

Keeping in mind the potential amount of new carbon tax revenue, its use and distribution could have appreciable impacts. In general, Congress would face a zero-sum game when deciding how to apportion the tax revenues: One revenue use necessarily forgoes the opportunity to apply that level of revenue elsewhere. Analogous to emission allowance value allocation[147] in a cap-and-trade program, the distribution of tax revenue would present policymakers with a series of tradeoffs. The primary options involve (1) minimizing the overall costs of the carbon tax imposed on society; (2) alleviating the costs borne by subgroups in society and economic sectors; and (3) providing funding to support other policy objectives, such as technological development. In the short-term, these options are mutually exclusive. However, in the long-term, advancements in low-carbon technology have the potential to reduce societal costs.

Overall costs to society

Overall costs may be referred to as macroeconomic costs and are often measured in terms of changes in projected gross domestic product (GDP) or another societal-scale metric—e.g., efficiency cost or welfare changes. Economic studies indicate that using carbon tax or auction revenues to offset reductions in distortionary taxes (labor, income) would be the most economically efficient use of the revenues and yield the greatest benefit to the economy overall. These studies show that if the revenues are used for other purposes, economic efficiency could suffer and the overall cost of the program would be higher.

A frequently cited study[148] compares the macroeconomic effects of different tax (or allowance auction) revenue applications with different emission allowance allocations in a cap-and-trade program. The study concludes that the least economically efficient method of emission allowance distribution would be to provide all of the emission allowances to covered sources at no cost. This is illustrated as the right-hand (red) column in **Figure 9**. This study finds that efficiency costs could be reduced by 21% from this scenario if policymakers were to distribute carbon tax revenues to U.S. households in "lump-sum" payments—e.g., increase the standard tax deduction or provide payments to households (represented by the middle/blue column in **Figure 9**). The lowest overall costs of those modeled would be achieved by using carbon tax revenues to reduce taxes on labor or investment. This option is shown in the left/green column. Compared to the least efficient option—no-cost distribution in a cap-and-trade program—this option would reduce societal costs by 47%. **Figure 9** indicates this option also produces substantially lower (approximately 32%) costs than lump-sum distribution. This is explained by the "double- dividend" effect, described above. However, the next section points out that the distributional impacts of these revenue applications are very different.

Distributional impacts

In isolation, a carbon tax is regressive. This means that lower-income households would see a greater percentage increase of household income required for carbon tax-related costs—higher energy and/or electricity bills, primarily—than higher income households. A 2007 report estimated that households in the lowest income bracket could face increased costs equal to 3.4% of their household income; households in the highest bracket could see a cost increase of 0.8% (**Table 4**).[149]

If Congress were to redistribute tax revenues to households in lump-sum payments, the 2007 report estimates the impacts to household income would be more progressive: The lowest income brackets would see net gains, while higher income brackets would break even or experience relatively minor losses (**Table 4**).

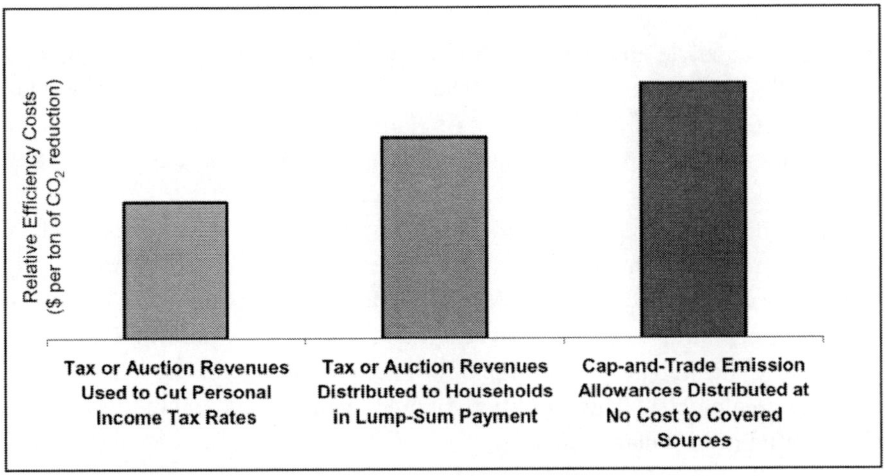

Source: Prepared by CRS with data from Lawrence H. Goulder, Mitigating the Adverse Impacts of CO2 Abatement Policies on Energy-Intensive Industries (2002), Resources for the Future Discussion Paper.

Figure 9. Relative Differences in Efficiency Costs between Different Applications of Tax (or Auction) Revenues and No-Cost Allowance Distribution in a Cap-and-Trade Program

Table 4. Distributional Effects of Carbon Tax with Different Applications of Carbon Tax Revenues

	Percentage Impact to Household Income by Income Bracket (1= Lowest, 10=Highest)									
Distributional Scenarios	**1**	**2**	**3**	**4**	**5**	**6**	**7**	**8**	**9**	**10**
Carbon Tax in Isolation (before revenue recycling)	-3.4	-3.1	-2.4	-2.0	-1.8	-1.5	-1.4	-1.2	-1.1	-0.8
Carbon Tax with a Lump-Sum Distribution to Households	2.1	1.0	0.6	0.4	0.3	0.1	-0.1	-0.1	-0.2	-0.2
Carbon Tax with Payroll Tax Rebate for Workers	-0.7	-1.0	-0.2	0.1	0.1	0.3	0.2	0.2	0.0	0.0
Carbon Tax with Payroll Tax Rebate for Workers and Equivalent Rebate for Social	1.4	1.0	0.6	0.3	0.1	0.1	0.1	-0.1	-0.1	-0.2

Source: Prepared by CRS with data from the following: Gilbert Metcalf, A Proposal for a U.S. Carbon Tax Swap: An Equitable Tax Reform to Address Global Climate Change (2007), The Hamilton Project, Brookings Institution.

Note: These results do not account for consumers' behavioral responses to the carbon tax. Metcalf points out that the results provide a "reasonable first approximation" of the different impacts to household incomes (Metcalf (2008)).

In another scenario examined by the 2007 study, workers would receive a tax rebate equal to the payroll taxes paid—including worker and employer contributions—in a given year.[150] The rebate would be capped at the same level ($560/year) for each worker. As **Table 4** indicates, this option, while alleviating much of the regressiveness of the carbon tax, would disproportionately impact the two lowest income brackets.[151]

This disproportionate outcome would be expected, because some of the lower income households may contain unemployed or retired persons, who would not benefit from a payroll tax rebate but still experience the higher energy prices brought on by the carbon tax. The last scenario in **Table 4**, in addition to the rebates for workers, would provide equal rebates to social security recipients. This addition would generate a more progressive distribution of tax-related impacts.

Carbon-intensive industries

This application of tax revenue would likely generate intense debate from industry and non- industry stakeholders alike. In general, stakeholders from all carbon-intensive industries argue that their businesses would be disproportionately affected by a carbon tax and thus should receive a portion of the revenues to avoid financial losses or provide transition into the new carbon- constrained regime. One counter-argument is that revenues used for this purpose would be unavailable to support other objectives (some of which are described in this chapter), but that concern applies to any use of the tax revenues. Regardless, many would assert that providing some level of assistance to carbon-intensive industries may be a political necessity.[152]

Complicating the debate is the fact that the impacts of a carbon tax would vary across and within carbon-intensive industries. In the context of carbon tax impacts, there are two broad categories of carbon-intensive industries: industrial sectors that can more easily pass along tax-related costs and those that cannot. The first group would likely include fossil fuel producers and the electric utility sector. Fossil fuel producers and electricity generating facilities are generally expected to pass along the vast majority of their tax-related costs. Some contend that providing assistance to these industries would run counter to the principle of creating a level playing field for low-carbon energy sources. In addition, there are likely to be both winners and losers within the electricity generation sector. For instance, some facilities, such as coal-fired plants, are expected to see greater losses, while others—hydroelectric or renewable energy plants—may see financial gains under a carbon tax or cap-and-trade program.[153]

The second category of carbon-intensive industries includes economic sectors that may face a comparative disadvantage if they were subject to GHG controls (directly or indirectly) and their international competition were not.[154] Because they may be constrained in their ability to pass along tax-related costs, there may be a stronger argument for providing assistance to this subset of carbon-intensive industries. However, it may be difficult for policymakers to determine the relative capacities of different industries and companies to pass along costs, or the degree of competitive disadvantage borne.[155]

Policymakers have other options—e.g., to attach a carbon price to imported materials—that could address potential comparative disadvantages that may arise under a carbon tax system.[156]

Recent legislative proposals would have provided some assistance to carbon-intensive industries. In the 110th Congress, Representative Larson's carbon tax proposal (H.R. 3416) would have allotted to "negatively affected" industries[157] approximately 8% of the tax revenues in the first year of the tax (2008 as written), declining to zero by 2018.[158] In addition, all of the cap-and-trade bills proposed in the 110th Congress that specified an emission allowance distribution strategy (as opposed to delegating the decision to a federal agency) would have allotted some percentage of allowance value to carbon-intensive industries.[159]

Technology development

Many experts consider government investment in technology an essential component[160] of a robust strategy to mitigate GHG emissions, because long-term emission reduction would require major restructuring of the energy economy. Moreover, the development and deployment of low- carbon technologies could have a considerable impact on the overall costs of an emissions reduction program. This is particularly the case if revenues induced technologies to be commercialized ahead of their projected schedules or if unanticipated low- or zero-carbon alternatives can be developed.

Although dispensing tax revenues for technological development could yield considerable gain, the return in terms of additional investment to promote technology improvements remains uncertain.[161] For example, some programs may be operating at full capacity, and additional funding may not provide comparable benefits. Moreover, the marketplace already provides some incentive for technological change, and a carbon tax would increase the existing incentive to develop low-carbon or zero-carbon energy alternatives.[162] In addition, using federal funds to support technology raises a number of issues for policymakers that are beyond the scope of this chapter.[163]

CONCLUSIONS

Market-based mechanisms that limit greenhouse gas (GHG) emissions can be divided into two types: quantity control (e.g., cap-and-trade) and price control (e.g., carbon tax). To some extent, a carbon tax and cap-and-trade program would produce similar effects: Both would place a price on carbon, and both are estimated to increase the price of fossil fuels, which would ultimately be borne by energy users, particularly households.

In an economically efficient market with perfect information, either a price (carbon tax) or quantity control instrument (cap-and-trade system) could be designed to achieve the same outcome. Because this market ideal does not exist, preference for a carbon tax or a cap-and-trade program ultimately depends on which variable one prefers to control—emissions or costs. Although there are several design mechanisms that could be included with either program that would blur the distinction, the gap between price control and quantity control can never be completely overcome.

Compared to a cap-and-trade system, a carbon tax may both offer several advantages as well as present several possible disadvantages. A known price on carbon would not add to the volatility of energy prices, such as electricity and household fuels, unlike what may occur under a cap-andtrade program. Moreover, with a set price, industry would have better information to guide investment decisions (e.g., energy efficiency improvements and/or equipment upgrades), unless the tax rate was subject to periodic adjustment to achieve GHG emission reduction goals. Economists often highlight the relative economic efficiency advantage of a carbon tax, but this advantage rests on assumptions about the expected costs and benefits of climate change mitigation, both of which contain considerable uncertainty and some controversy. Some contend that a carbon tax may provide implementation advantages, including greater transparency, reduced administrative burden, and relative ease of modification. In addition, some argue that a carbon tax approach may make it easier to reach a GHG reduction agreement with developing nations. However, some counter that a cap-and-trade approach is best suited to meet this objective.

The primary disadvantage of a carbon tax is that it would yield uncertain emissions. For some this concern may present a non-starter, precluding a carbon tax as an option to control GHG emissions. Some argue that the potential for irreversible climate change impacts necessitates the emissions certainty that is only available with a quantity-based instrument (e.g., cap-and-trade). However, depending upon its design, a cap-and-trade program may also include a risk of uncertain emissions. For example, if offsets are allowed as a compliance option, there is some concern that they would not represent real emission reductions. Emissions leakage offers another threat to the certainty of the emissions cap.

Although it may present implementation challenges, policymakers could devise a tax program that yields only short-term emission fluctuations, as it progresses towards its long-term emission reduction objective. Proponents argue that short-term emission fluctuations would be preferable to price fluctuations that might be expected with a cap-and-trade system.

Many assert that it would be easier to garner support for a cap-and-trade system, because a tax increase policy typically carries a political stigma. To some extent, this argument may be viewed by some as based on semantics, because a cap-and-trade program would act very much like a tax, in terms of its effects on energy prices. However, a carbon tax and a cap-and-trade legislation would fall into different committee jurisdictions within Congress. The committees developing the GHG emission legislation may have Members with varying perceptions about the relative values of uncertain costs versus uncertain emission reductions. The committee jurisdiction factor may play a role in the debate over the preferred policy instrument.

In addition, some environmental groups may oppose a carbon tax, because they assume the tax rate would be set too low to be effective in controlling GHG emissions. This is more of an argument over stringency than over the instrument used to mitigate GHG emissions.

The debate over a carbon tax versus a cap-and-trade program may be instructive for policymakers because it highlights (1) the primary differences between the instruments and the (2) design elements that could be included within either instrument that would blur their distinctions. For example, Congress could devise a cap-and-trade program that controls total costs and price volatility to some degree. However, a cap-and-trade price control mechanism, such as a safety- valve and/or reserve price, would alter the character of the cap-and-trade program—the level of emissions would no longer be certain. This potential outcome reinforces the fundamental choice between a carbon tax and a cap-and-trade program: cost certainty versus emissions certainty.

APPENDIX. ADDITIONAL INFORMATION

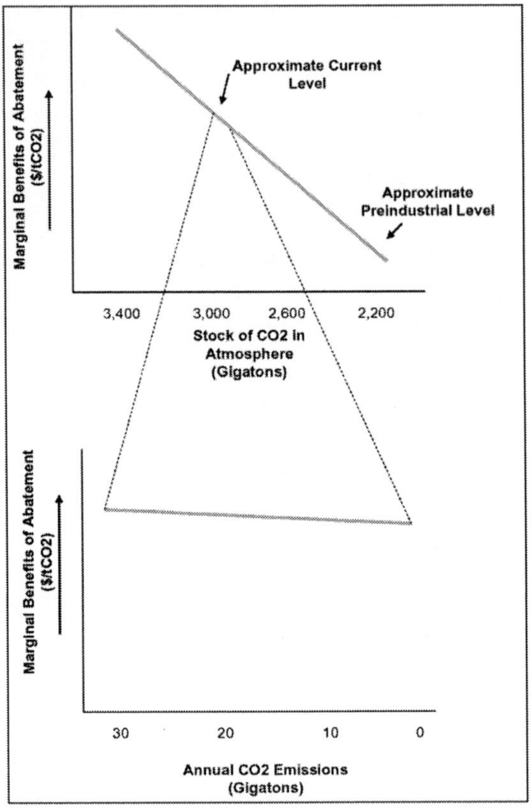

Source: Prepared by CRS. Adapted from William Pizer, *Prices vs. Quantities Revisited: The Case of Climate Change* (1997), Resources for the Future Discussion Paper 98-02.

Note: This figure illustrates the concept that for a stock pollutant, such as CO_2, reductions made in one year would have minimal effect on the pollutant's stock. As shown at the top of the figure, the current stock of CO_2 is approximately 3,000 gigatons (1 gigaton = 1 billion tons). Human-related activities generated approximately 27 gigatons of CO_2 emissions in 2005. If all nations were to stop emitting CO_2 emissions in one year, the effect on the stock of CO_2 would likely be minimal. Thus, the marginal benefit of abatement (for one year) is thought to be relatively flat, as depicted in the bottom of the figure. It would take multiple years of zero emissions to impact CO_2 emissions.

Figure A-1. Illustration of Relationship between the Stock of CO_2 in Atmosphere and Annual CO_2 Emissions

Table A-1. Comparison of Estimated Carbon Tax-Related Price Impacts to Fossil Fuels and Motor Gasoline from Selected Carbon Tax Rates

Carbon Tax-Related Price Impacts								
Carbon Tax Rate	**Coal (short ton)**		**Oil (barrel)**		**Natural Gas (thousand cubic feet, mcf)**		**Motor Gasoline (gallon)**	
\$5/ $mtCO_2$	\$9.45/ short ton	28%	\$2.15/barrel	5%	\$0.28/ mcf	2%	\$0.05/gallon	2%
\$15/$mtCO_2$	\$28.35/ short ton	83%	\$6.45/barrel	16%	\$0.83/ mcf	6%	\$0.14/gallon	4%
\$25/$mtCO_2$	\$47.25/ short ton	138%	\$10.75/barrel	27%	\$1.38/ mcf	10%	\$0.23/gallon	7%
\$50/$mtCO_2$	\$94.50/ short ton	276%	\$21.50/barrel	55%	\$2.75/ mcf	20%	\$0.45/gallon	14%
Inputs Used in Above Comparison[a]								
Metric tons of CO_2 ($mtCO_2$) per unit of fuel[b]	1.89 $mtCO_2$/ short ton[c]		0.43 $mtCO_2$/ barrel[d]		0.055 $mtCO_2$/ mcf[e]		0.009 $mtCO_2$/ gallon[f]	

a. These estimates assume that the carbon tax is applied upstream to fossil fuels. The price impacts assume that the carbon tax is passed through to end-users and not absorbed by upstream energy producers. The price effects also do not take into account behavioral changes—e.g., energy conservation, fuel substitution, etc.—in response to price increases.

b. Values derived from CO_2 coefficients (i.e., CO_2 emissions per quadrillion BTU) and thermal conversion factors (i.e., BTU per fuel unit)for each fuel. CO_2 coefficients are from EIA, *Documentation for Emissions of Greenhouse Gases in the United States 2006* (2008), table 6-1; thermal conversion factors from EIA, Annual Energy Review 2007 (2008), Appendix A.

c. This value represents the CO_2 coefficient for coal (electric power sector) and the thermal conversion factor for coal consumption from the electric power sector (2007).

d. This value represents the CO_2 coefficient for crude oil and the thermal conversion factor for "unfinished oil."

e. This value represents CO_2 coefficient for natural gas (pipeline) and the thermal conversion factor for endues sectors (2007).

f. This value represents the CO_2 coefficient for "motor gasoline" and the thermal conversion factor for motor gasoline (conventional).

End Notes

[1] The major GHGs discussed include carbon dioxide (CO2), methane (CH4), nitrous oxide (N2O), sulfur hexafluoride (SF6), hydrofluorocarbons (HFC), and perfluorocarbons (PFC). Recent GHG reduction proposals have also included nitrogen trifluoride (NF3).

[2] As discussed in this chapter, terminology is a key issue. Some proponents of the "carbon tax" approach describe the policy instrument as a user fee or user charge. There are multiple reasons that proponents may seek to change the nomenclature. Perhaps the primary concern is the political stigma associated with the word "tax." Regardless, the term "carbon tax" is the one that is most commonly associated with the GHG control policy instrument discussed in this report.

[3] Non-carbon GHGs could still be subject to the tax based on their contribution to global warming in relation to CO2. Global warming potential (GWP) is an index of how much a GHG may contribute to global warming over a period of time, typically 100 years. GWPs are used to compare gases to carbon dioxide, which has a GWP of 1. For example, methane's GWP is 25, and is thus 25 times more potent a GHG than CO2. The GWPs listed in this report are from: Intergovernmental Panel on Climate Change, *Climate Change 2007: The Physical Science Basis* (2007), p. 212.

[4] See e.g., Gilbert Metcalf, *A Green Employment Tax Swap: Using a Carbon Tax to Finance Payroll Tax Relief* (2007); Nathaniel Keohane and Sheila Olmstead, *Markets and the Environment* (2007), Island Press; Ian Parry "Fiscal Interactions and the Case for Carbon Taxes over Grandfathered Carbon Permits," in *Climate Change Policy* (Dieter Helm, editor), Oxford University Press (2005).

[5] Named after A. Cecil Pigou, author of a landmark economic work, *Economics of Welfare* (1920).

[6] See also, CRS Report RL33799, *Climate Change: Design Approaches for a Greenhouse Gas Reduction Program*, by Larry Parker; CRS Report RL34513, *Climate Change: Current Issues and Policy Tools*, by Jane A. Leggett.

[7] See e.g., William Pizer, *Prices vs. Quantities Revisited: The Case of Climate Change* (1997), Resources for the Future Discussion Paper 98-02.

[8] The marginal cost curve indicates the cost of an *additional* unit (e.g., one ton of GHG emissions) of emission abatement at different emission levels. A rising marginal cost curve (as depicted in **Figure 1**) signals that as each incremental unit of GHG emission abatement is made, the cost of abatement (per unit) increases.

[9] The marginal benefit curve indicates the benefit of an *additional* unit (e.g., one ton of GHG emissions) of emission abatement at different emission levels. A declining benefit slope (as depicted in **Figure 1**) illustrates that with each incremental unit of abatement, the per unit benefit decreases.

[10] Abatement may include emission reductions, emissions avoided, and sequestration activities.

[11] In other words, for abatement above this level, it would cost more than the value of the benefits received. Society would be paying too much; money would be better spent for other purposes.

[12] See CRS Report RL33 970, *Greenhouse Gas Emission Drivers: Population, Economic Development and Growth, and Energy Use*, by John Blodgett and Larry Parker.

[13] For a analysis of some recent assessments, see CRS Report RL34489, *Climate Change: Costs and Benefits of S. 2191/S. 3036*, by Larry Parker and Brent D. Yacobucci.

[14] See also, CRS Report RL33799, *Climate Change: Design Approaches for a Greenhouse Gas Reduction Program*, by Larry Parker.

[15] Emission allowances are essentially the currency in a cap-and-trade program. An allowance is generally defined as a limited authorization by the government to emit 1 ton of pollutant. In the case of GHGs, an allowance generally refers to a metric ton of carbon dioxide equivalent (tCO2-e).

[16] A 2008 Resources for the Future (RFF) report estimated the effects banking provisions had on price volatility. The authors concluded that banking eliminates about 20% of the cost differences between a price control (tax) system and cap-and-trade without banking provisions. Harrison Fell *et al.*, *Prices Versus Quantities Versus Bankable Quantities* (2008), Resources for the Future Discussion Paper.

[17] See CRS Report RL34502, *Emission Allowance Allocation in a Cap-and-Trade Program: Options and Considerations*, by Jonathan L. Ramseur.

[18] See Dieter Helm *et al.*, "Credible Carbon Policy," in *Climate Change Policy* (ed. Dieter Helm) (2005).

[19] M. L. Weitzman, "Price vs. Quantities" (1974), *Review of Economic Studies*, 41(4): 477-491.

[20] This conclusion is often referred to as the "Weitzman Rule." Nathaniel Keohane and Sheila Olmstead, *Markets and the Environment* (2007), Island Press

[21] In other words, not all emission reductions (or sequestration activities) are equal in terms of cost: there may be "low- hanging fruit" opportunities as well as actions that would require more substantial capital investments.

[22] Richard Newell and William Pizer, "Regulating Stock Externalities under Uncertainty," *Journal of Environmental Economics and Management* 45: 4 16-432 (2003); Warwick J. McKibbin and Peter J. Wilcoxen, "The Role of Economics in Climate Change Policy," (2002) *Journal of Economic Perspectives* 16(2): 107-29.

[23] The "curves" in this report's figures are depicted as straight lines for illustrative purposes.

[24] Although marginal benefit information is also uncertain, the relative-slope principle generally depends only on the marginal cost curve. The reasoning behind this approach and circumstances that may undermine this approach are discussed below.

[25] Other GHG emissions may yield damages unrelated to climate change impacts.

[26] 1 gigaton = 1 billion tons.

[27] William Pizer, *Prices vs. Quantities Revisited: The Case of Climate Change* (1997), Resources for the Future Discussion Paper 98-02.

[28] See e.g., James Hansen, Testimony before the House Select Committee on Energy Independence and Global Warming (June 20, 2008).

[29] See Timothy M. Lenton, *et al.*, "Tipping Elements in the Earth's Climate System," *Proceedings of the National Academy of Sciences* 105:6 (February 2008).

[30] Nicholas Stern, *The Economics of Climate Change: The Stern Review* (2006), p. 314 (emphasis in original).

[31] Nicholas Stern, *The Economics of Climate Change: The Stern Review* (2006), p. 314, footnote 12.

[32] William Pizer, *Climate Change Catastrophes* (2003), Resources for the Future Discussion Paper. See also, CBO, *Policy Options for Reducing CO2 Emissions* (2008).

[33] See CRS Report RL34489, *Climate Change: Costs and Benefits of S. 2191/S. 3036*, by Larry Parker and Brent D. Yacobucci.

[34] CBO, *Policy Options for Reducing CO2 Emissions* (2008).

[35] William Pizer, *Prices vs. Quantities Revisited: The Case of Climate Change* (1997), Resources for the Future Discussion Paper 98-02.

[36] Lawrence Goulder and Ian Parry, *Instrument Choice in Environmental Policy* (2008), Resources for the Future Discussion Paper, implicitly referencing: Michael Hoel and Larry Karp, "Taxes Versus Quotas for a Stock Pollutant," *Resource and Energy Economics* (2002) 24:367-84; Newell and Pizer (2003); and Larry Karp and Jiangfeng Zhang, "Regulation of Stock Externalities with Correlated Abatement Costs," *Environmental and Resource Economics* (2005) 32:273-299.

[37] Robert Stavins, "Correlated Uncertainty and Policy Instrument Choice," *Journal of Environmental Economics and Management* 30:218-232 (1996).

[38] Newell and Pizer (2003).

[39] Gilbert Metcalf *et al.*, *Analysis of U.S. Greenhouse Gas Tax Proposals* (2008), MIT Joint Program on the Science and Policy of Climate Change, Report No. 160.

[40] Pizer (1997).

[41] See, Hoel and Karp (2002); Newell and Pizer (2003); Karp and Zhang (2005).

[42] Ibid.

[43] Metcalf *et al.* (2008).

[44] Harrison Fell *et al.*, *Prices Versus Quantities Versus Bankable Quantities* (2008), Resource for the Future Discussion Paper.

[45] See Alan Ingham and Alistair Ulph, "Uncertainty and Climate Change Policy,"in *Climate Change Policy* (Dieter Helm, editor), Oxford University Press (2005).

[46] Indeed, the Carbon Tax Center lists price volatility control as one of the fundamental reasons taxes are superior to emissions caps. Carbon Tax Center, Tax vs. Cap-and-Trade, at [www.carbontax.org].

[47] William Nordhaus, "To Tax or Not to Tax: Alternative Approaches to Slowing Global Warming" (2007), *Review of Environmental Economics and Policy*, Vol. 1 (1): 26-44.

[48] Energy Information Administration, *Annual Energy Outlook 2008* (2008).

[49] Nordhaus (2007).

[50] Ian W.H. Parry and William Pizer, *Emissions Trading Versus CO2 Taxes Versus Standards* (2007), Resources for the Future Issue Brief.

[51] See CRS Report RL34150, *Climate Change and the EU Emissions Trading Scheme (ETS): Kyoto and Beyond*, by Larry Parker.

[52] Although Phase 2 began in 2008, the market for Phase 2 allowances was created in 2005. Unlike Phase 1 allowances, Phase 2 allowances may be banked for future use in a planned Phase 3 anticipated to begin in 2013.

[53] See CRS Report RL34502, *Emission Allowance Allocation in a Cap-and-Trade Program: Options and Considerations*, by Jonathan L. Ramseur.

[54] Carbon Tax Center, "Tax vs. Cap-and-Trade," at http://www.carbontax.org/issues/carbon

[55] See CRS Report RL34488, *Regulating a Carbon Market: Issues Raised By the European Carbon and U.S. Sulfur Dioxide Allowance Markets*, by Mark Jickling and Larry Parker.

[56] See CRS Report RL34436, *The Role of Offsets in a Greenhouse Gas Emissions Cap-and-Trade Program: Potential Benefits and Concerns*, by Jonathan L. Ramseur.

[57] See CRS Report RL34489, *Climate Change: Costs and Benefits of S. 2191/S. 3036*, by Larry Parker and Brent D. Yacobucci.

[58] Metcalf (2008).

[59] AEI, 2007.

[60] Parry and Pizer, 2007.

[61] As with other comparisons, this would depend upon the design of the cap-and-trade program.

[62] See Kenneth Green, *et al.*, *Climate Change: Caps vs. Taxes* (2007), American Enterprise Institute.

[63] *Ibid.*

[64] This potential problem could be alleviated by allowing covered sources to bank emission allowances.

[65] A cap-and-trade program, depending upon its design, may carry some risk of uncertain emissions as well. For example, if offsets are allowed as a compliance option, there is some concern that they would not represent real emission reductions. Emissions leakage is another threat to the certainty of the emissions cap. See CRS Report RL34436, *The Role of Offsets in a Greenhouse Gas Emissions Cap-and-Trade Program: Potential Benefits and Concerns*, by Jonathan L. Ramseur; and CRS Report R40100, *"Carbon Leakage" and Trade: Issues and Approaches*, by Larry Parker and John Blodgett.

[66] See e.g., Metcalf (2008) and Shapiro (2008).

[67] See Dieter Helm *et al.*, "Credible Carbon Policy," in *Climate Change Policy* (ed. Dieter Helm) (2005).

[68] Council on Foreign Relations, *Confronting Climate Change: A Strategy for U.S. Foreign Policy* (2008), Independent Task Force Report No. 161.

[69] See Letter from Representatives Sensenbrenner and Cantor to Representatives Conyers and Nadler (February 11, 2009), at http://www.eenews.net/features/documents/2009/02/11/document_pm_03.pdf. This letter also argues that if a cap-and-trade bill were to raise revenue, the bill should originate in the House of Representatives.

[70] See e.g., Gilbert Metcalf, *A Green Employment Tax Swap: Using a Carbon Tax to Finance Payroll Tax Relief* (2007).

[71] For example, the U.S. Climate Action Partnership (USCAP), a group of large companies (and environmental organizations) recommends that a "significant portion of allowances should be initially distributed free to capped entities and to economic sectors particularly disadvantaged by the secondary price effects of a cap.... " U.S. CAP, *A Call to Action*, at http://www.us-cap.org/.

[72] See CRS Report RL34150, *Climate Change and the EU Emissions Trading Scheme (ETS): Kyoto and Beyond*, by Larry Parker.

[73] See IPA Energy Consulting, *Implications of the EU Emissions Trading Scheme for the UK Power Generation Sector* (2005), Prepared for the United Kingdom Department of Trade and Industry; Jos Sijm et al., "CO2 Cost Pass-Through and Windfall Profits in the Power Sector," *Climate Policy* 6 (2006): 49-72; Point Carbon Advisory Services, *EU ETS Phase II—The Potential and Scale of Windfall Profits in the Power Sector* (2008), Prepared for World Wildlife Fund.

[74] See CRS Report RL33 846, *Greenhouse Gas Reduction: Cap-and-Trade Bills in the 110th Congress*, by Larry Parker, Brent D. Yacobucci, and Jonathan L. Ramseur.

[75] For example, the President and CEO of Duke Energy, who initially was an advocate a cap-and-trade approach (and remains a member of USCAP), described the Lieberman-Warner bill (S. 2191) as a "bastardized" version of cap-andtrade due to its allocation scheme. See Jim Rogers, "Climate Change Legislation Should Not Be Punitive" (February 2008), *Energy Daily*.

[76] Hepburn (2006), citing P.D. Klemperer, *Auctions: Theory and Practice* (2004), Princeton University Press.

[77] In a January 2009 speech, the CEO of ExxonMobil (Rex Tillerson) spoke in favor of a carbon tax over a cap-andtrade system, stating a carbon tax would be "the most efficient means of reflecting the cost of carbon in all economic decisions—from investments made by companies to fuel their requirements, to the product choices made by consumers." This quotation was taken from a UPI news report, at [www.upi.com].

[78] There may be some industry stakeholders who support a carbon tax on the assumption that it will be politically challenging for Congress to subsequently raise carbon taxes to the level needed to stimulate the stringent emissions reductions targeted (e.g., 80% below current levels by 2050) in recent cap-and-trade proposals.

[79] For example, different treatment of offsets could raise concern, particularly if covered entities in the United States gain access to questionable offset projects. See CRS Report RL34436, *The Role of Offsets in a Greenhouse Gas Emissions Cap-and-Trade Program: Potential Benefits and Concerns*, by Jonathan L. Ramseur.

[80] Qualified credits could range from EU emission allowances (EUAs) from the EU ETS to certified emission reductions (CERs)—i.e., international offsets—issued by the Clean Development Mechanism of the Kyoto Protocol.

[81] Scott Barrett and Robert Stavins, "Increasing Participation and Compliance in International Climate Change Agreements," *International Environmental Agreements: Politics, Law and Economics*, Vol. 3:349-76 (2003).

[82] However, Kyoto participants account for the majority of 1990 emissions (the Kyoto baseline).

[83] Data from World Resources Institute, Climate Analysis Indicators Tool (CAIT), at http://cait.wri.org/.

[84] See CRS Report RL34150, *Climate Change and the EU Emissions Trading Scheme (ETS): Kyoto and Beyond*, by Larry Parker; and CRS Report RL33826, *Climate Change: The*

Kyoto Protocol, Bali "Action Plan," and International Actions, by Susan R. Fletcher and Larry Parker.

[85] William Nordhaus and Joseph Boyer, *Warming the World: Economic Modeling of Global Warming* (2000), MIT Press.

[86] William Nordhaus, "To Tax or Not to Tax: Alternative Approaches to Slowing Global Warming" (2007), *Review of Environmental Economics and Policy*, Vol. 1 (1): 26-44.

[87] David Victor, *The Collapse of the Kyoto Protocol and the Struggle to Slow Global Warming* (2001), Princeton University Press.

[88] EPA, *Tools of the Trade: A Guide to Designing and Operating a Cap and Trade Program for Pollution Control* (2003), p. 3-5.

[89] For additional reading on these considerations, see e.g., Gilbert Metcalf and David Weisbach, *The Design of a Carbon Tax* (June 2008), Tufts University and the University of Chicago; EPA, *Tools of the Trade* (2003).

[90] Non-CO_2 GHG emissions come from a wide range of sources, most of which are outside the energy sector.

[91] Power plants may be considered midstream, because their emissions are influenced by the demand from electricity consumers: businesses and households. In addition, some carbon-intensive industries, such as steel or cement, might be categorized as midstream.

[92] Section 821 of the 1990 Clean Air Act Amendments requires electric generating facilities affected by the acid rain provisions of Title IV to monitor CO_2 in accordance with Environmental Protection Agency (EPA) regulations.

[93] Although a price signal would be sent in either case—thus (in theory) encouraging CCS development—some contend that if coal is taxed upstream at the extraction point, some of the price signal may be weakened before it reaches coal-fired emission sources.

[94] For more information on this technology, see CRS Report RL33801, *Carbon Capture and Sequestration (CCS)*, by Peter Folger.

[95] See e.g., John Reilly *et al.*, "The Role of Non-CO_2 GHGs in Climate Policy: Analysis Using the MIT IGSM" (2006), *The Energy Journal*, Special Issue No. 3: 503-520.

[96] Metcalf *et al.* (2008). The overall costs fall, because the tax rate is decreased. By including all GHGs, the tax rate can be lowered if the cumulative target remains the same.

[97] Gilbert Metcalf and David Weisbach, *The Design of a Carbon Tax* (June 2008), Tufts University and the University of Chicago.

[98] There are additional GHGs, but they are not included in the current U.S. GHG inventory. For more information, see CRS Report RL34266, *Climate Change: Science Highlights*, by Jane A. Leggett.

[99] Based on 2006 data. EPA, *Inventory of U.S. Greenhouse Gas Emissions and Sinks: 1990-2006* (April 2008).

[100] For example, a 5-cent-per-barrel tax on the oil industry—domestically produced oil and oil imports—supports the Oil Spill Liability Trust Fund.

[101] In 2006, there were 3,292 plants that had at least one fossil fuel-fired generator (EIA, 860 data files).

[102] EIA, *Annual Coal Report 2006* (2007).

[103] Industry used 5% of the coal in 2006; residences and commercial entities used 3%.

[104] EPA, *Inventory of U.S. Greenhouse Gas Emissions and Sinks: 1990-2006* (April 2008).

[105] Metcalf and Weisbach (2008).

[106] Some of the cap-and-trade proposals made in recent years would delegate coverage decisions to the implementing agency, typically EPA.

[107] Alternatively, policymakers could develop emission/sequestration standards for specific activities. This latter option would not be a market-based approach, but it was included in one of the cap-and-trade proposals from the 1 10th Congress—H.R. 6186 (Markey).

[108] See EPA's Landfill Methane Outreach Program, at http://www.epa.gov/lmop/overview.htm.

[109] A small amount of perfluorocarbons (PFCs) are also used as substitutes (EPA Inventory, 2008). Under the Montreal Protocol and the Clean Air Act Amendments of 1990, ODSs are

being phased out. See CRS Report RL30853, *Clean Air Act: A Summary of the Act and Its Major Requirements*, by James E. McCarthy et al.

[110] EPA Inventory (2008).

[111] See e.g., Metcalf and Weisbach (2008).

[112] Intergovernmental Panel on Climate Change, *Safeguarding the Ozone Layer and the Global Climate System, Issues related to Hydrofluorocarbons and Perfluorocarbons* (2005).

[113] As with fossil fuels, policymakers may want to consider applying the tax to imported steel and cement. This would increase the number of entities subject to the tax.

[114] See e.g., Metcalf and Weisbach (2008).

[115] EPA Inventory (2008).

[116] EPA Inventory (2008).

[117] A comprehensive discussion and analysis of climate change costs and benefits is beyond the scope of this report. This section highlights issues associated with estimating and applying climate change cost and benefit estimates for carbon tax implementation.

[118] IPCC, *Climate Change 2007: Mitigation*, (Annex I—Glossary)Contribution of Working Group III to the Fourth Assessment Report of the Intergovernmental Panel on Climate Change (2007).

[119] Richard Tol, "The Social Cost of Carbon: Trends, Outliers, and Catastrophes," *Economics E-Journal* (August 2008). This updates a previous study: Richard Tol, "The Marginal Damage Costs of Carbon Dioxide Emissions: An Assessment of the Uncertainties," *Energy Policy* 33:2064-2074 (2005)

[120] See e.g., Nicholas Stern, *The Economics of Climate Change: The Stern Review* (2006); Richard Tol, "The Marginal Damage Costs of Carbon Dioxide Emissions: An Assessment of the Uncertainties," *Energy Policy* 33:2064-2074 (2005); David Pearce, "The Social Cost of Carbon," in *Climate Change Policy* (Dieter Helm, editor), Oxford University Press (2005).

[121] For a more comprehensive discussion and explanation of discount rates and their applicability to climate change, see David Weisbach and Cass Sunstein, *Climate Change and Discounting the Future: A Guide for the Perplexed*, Reg- Markets Center Working Paper No. 08-19 (August 2008); and Jiehan Guo *et al.*, "Discounting and the Social Cost of Carbon: A Closer Look at Uncertainty" *Environmental Science and Policy*, 9: 205-126 (2006).

[122] Intergovernmental Panel on Climate Change, *Climate Change 2007: Impacts, Adaptation, and Vulnerability* (2007), Working Group II Contribution to the Fourth Assessment Report, p. 823.

[123] Nicholas Stern, *The Economics of Climate Change: The Stern Review* (2006), p. 26.

[124] Stern's business-as-usual estimate for the SCC was $95 per metric ton of CO2-e (in $2005); the SCC estimate for a pathway that would stabilize GHG concentration at 450 parts per million was $28.

[125] Ibid., .., p. 301.

[126] Gilbert Metcalf, *A Proposal for a U.S. Carbon Tax Swap: An Equitable Tax Reform to Address Global Climate Change* (2007), The Hamilton Project, Brookings Institution.

[127] Nicholas Stern, *The Economics of Climate Change: The Stern Review* (2006), p. 301.

[128] See Robert Shapiro *et al.*, *Addressing Climate Change Without Impairing the U.S. Economy: The Economics and Environmental Science of Combining a Carbon-Based Tax and Tax Relief* (2008), U.S. Climate Task Force; and Gilbert Metcalf, *A Proposal for a U.S. Carbon Tax Swap: An Equitable Tax Reform to Address Global Climate Change* (2007), The Hamilton Project, Brookings Institution.

[129] Richard Tol, "The Social Cost of Carbon: Trends, Outliers, and Catastrophes," *Economics E-Journal* (August 2008).

[130] If non-peer-reviewed estimates were included, the average would be much higher, because these estimates are generally in the upper end of the range. See Richard Tol, "The Social Cost of Carbon: Trends, Outliers, and Catastrophes," *Economics E-Journal* (August 2008).

131 Intergovernmental Panel on Climate Change, *Climate Change 2007: Impacts, Adaptation, and Vulnerability* (2007), Working Group II Contribution to the Fourth Assessment Report.

132 For example, William Nordhaus estimates a SCC of approximately $8/tCO2-e (*A Question of Balance: Economic Modeling of Global Warming* (2008), prepublication version,Yale Press); Nicholas Stern's SCC estimate is approximately $26/tCO2-e. Both of these estimates are in $2005 and are for different business-as-usual scenarios (Nicholas Stern, *The Economics of Climate Change: The Stern Review* (2006)).

133 This is the projected emission reduction from the Lieberman-Warner proposal (S. 2191/S. 3036).

134 Selecting a GHG reduction goal would be a complicated endeavor, involving multiple considerations. This discussion is beyond the scope of this report.

135 For a comprehensive analysis of these issues see CRS Report RL34489, *Climate Change: Costs and Benefits of S. 2191/S. 3036*, by Larry Parker and Brent D. Yacobucci.

136 Although some (primarily economists) may define this phrase more narrowly—i.e., only in the context of offsetting distortionary taxes—in this report, revenue recycling refers to any application of carbon tax revenue.

137 Neither bill would address other GHG emissions.

138 U.S. Internal Revenue Service, *Internal Revenue Service Data Book 2005* (2006)..

139 Ian W.H. Parry and William Pizer, *Emissions Trading Versus CO2 Taxes Versus Standards* (2007), Resources for the Future Issue Brief.

140 Parry, "Fiscal Interactions and the Case for Carbon Taxes over Grandfathered Carbon Permits," in *Climate Change Policy* (Dieter Helm, editor), Oxford University Press (2005); Intergovernmental Panel on Climate Change Working Group III, *Climate Change 2001: Mitigation*, (Cambridge, UK: Cambridge University Press, 2001) Chapter 8.

141 This is referred to as the "tax-interaction effect" in economic literature. In short, a carbon tax would increase energy costs. Increased energy costs are expected to lead to some decrease in real household income and labor supply. See e.g., Ian Parry, "Fiscal Interactions and the Case for Carbon Taxes over Grandfathered Carbon Permits," in *Climate Change Policy* (Dieter Helm, editor), Oxford University Press (2005)

142 There is some evidence that the double-dividend benefits are stronger in Europe than in the United States, because the former has a more stringent income tax system. With more stringent taxes in place, there is more "distortion" to reduce, thus strengthening the double-dividend effect. Intergovernmental Panel on Climate Change Working Group III, *Climate Change 2001: Mitigation*, (Cambridge, UK: Cambridge University Press, 2001) Chapter 8, citing several studies.

143 For cost estimates from economic models, see Metcalf *et al.* (2008) and Robert Shapiro *et al.*, *Addressing Climate Change Without Impairing the U.S. Economy: The Economics and Environmental Science of Combining a Carbon- Based Tax and Tax Relief* (2008), U.S. Climate Task Force.

144 For more discussion, see CRS Report R40100, *"Carbon Leakage" and Trade: Issues and Approaches*, by Larry Parker and John Blodgett.

145 See e.g., CBO, *Evaluating the Role of Prices and R&D in Reducing Carbon Dioxide Emissions*, (2006).

146 A carbon tax would raise energy costs, which is expected to lead to some reduction in real household wages and labor supply. In turn, federal tax revenue may decline.

147 Allowance value allocation includes both the use of auction revenues and distribution of allowances to entities (e.g., industry or non-covered parties) at no cost.

148 In this study, the author applies a carbon tax of approximately $7/tCO2 that rises (at 7% or 9%) until it reaches $14/tCO2, at which point the tax is held constant in real dollars. The study also examines an emissions trading program with allowance prices that match the carbon tax paths. Lawrence H. Goulder, *Mitigating the Adverse Impacts of CO2 Abatement Policies on Energy-Intensive Industries* (2002), Resources for the Future Discussion Paper.

[149] Gilbert Metcalf, *A Proposal for a U.S. Carbon Tax Swap: An Equitable Tax Reform to Address Global Climate Change* (2007), The Hamilton Project, Brookings Institution.

[150] The Social Security payroll tax applies a constant tax rate—half of which is paid by the employee and the other half paid by the employer (the self-employed are responsible for the entire amount)—up to a certain income level. According to the Tax Policy Center, for nearly two-thirds of U.S. households, payroll taxes (if the employer portion is counted) are the single largest tax paid to the federal government at: http://www.taxpolicycenter.org/publications/ url.cfm?ID=1001065.

[151] Because this option effectively replaces a portion of the payroll tax (a regressive tax on an economic good) with a carbon tax (a levy on an economic bad), it has the potential to improve economic efficiency, and thus lower overall society costs. However, Metcalf asserts that the rebate is unlikely to appreciably alter labor supply, except among low-income workers. See Metcalf (2007).

[152] For a further discussion of rationales for providing support to carbon-intensive industries, see CRS Report R40 100, *"Carbon Leakage" and Trade: Issues and Approaches*, by Larry Parker and John Blodgett.

[153] Dallas Burtraw and Karen Palmer, *Compensation Rules for Climate Policy in the Electricity Sector* (2007), Resources for the Future Discussion Paper.

[154] Among others, the steel, paper, and aluminum industries are often cited as examples of trade-exposed industries.

[155] Department of Climate Change, Commonwealth of Australia, *Carbon Pollution Reduction Scheme: Green Paper* (July 2008), p. 310.

[156] See CRS Report R40 100, *"Carbon Leakage" and Trade: Issues and Approaches*, by Larry Parker and John Blodgett.

[157] As determined by the Secretary of the Treasury in consultation with the Secretary of Labor.

[158] The other carbon tax proposal from the 110th Congress—H.R. 2069 (Stark)—does not include revenue distribution provisions.

[159] See CRS Report RL33846, *Greenhouse Gas Reduction: Cap-and-Trade Bills in the 110th Congress*, by Larry Parker, Brent D. Yacobucci, and Jonathan L. Ramseur.

[160] Some argue that the private sector is likely to under-invest in research and development, even with a carbon price in place. See W. David Montgomery and Anne E. Smith, *Price, Quantity, and Technology Strategies for Climate Change Policy* (2005), CRA International.

[161] See Robert W. Hahn, *Greenhouse Gas Auctions and Taxes: Some Practical Considerations* (2008), AEI Center for Regulatory and Market Studies.

[162] See CRS Report RL34621, *Capturing CO2 from Coal-Fired Power Plants: Challenges for a Comprehensive Strategy*, by Larry Parker, Deborah D. Stine, and Peter Folger.

[163] See CRS Report RL34513, *Climate Change: Current Issues and Policy Tools*, by Jane A. Leggett.

In: U.S. Energy and the Environment… ISBN: 978-1-61668-017-6
Editors: Roland H. Terrison pp.179-190

Chapter 7

ENVIRONMENTAL QUALITY INCENTIVES PROGRAM (EQIP): STATUS AND ISSUES

Megan Stubbs

SUMMARY

The Environmental Quality Incentives Program (EQIP) is a voluntary program that provides farmers with financial and technical assistance to plan and implement soil and water conservation practices. EQIP is the largest agriculture conservation financial assistance program for working lands. EQIP was first authorized in 1996 and was most recently revised by Section 2501 of the Food, Conservation, and Energy Act of 2008 (P.L. 110-246, the 2008 farm bill). It is a mandatory spending program (i.e., not subject to annual appropriations) and is administered by the U.S. Department of Agriculture's (USDA's) Natural Resources Conservation Service (NRCS). Funding is currently authorized to grow to $1.75 billion in FY20 12. Eligible land includes cropland, rangeland, pasture, non-industrial private forestland, and other land on which resource concerns related to agricultural production could be addressed through an EQIP contract.

With the 111th Congress facing tighter budget constraints, EQIP could face similar challenges with a potential reduction in mandatory funding levels and a continuing backlog of unfunded applications. A change in income limitations along with a new waiver created in the 2008 farm bill could also raise issues for the program. EQIP will also continue to face challenges in measuring environmental and program accomplishments.

PROGRAM OVERVIEW

The Environmental Quality Incentives Program (EQIP) is a voluntary program that provides technical and financial assistance to eligible agricultural producers who wish to implement soil and water conservation practices. The purpose of EQIP is to promote agriculture production, forestry management, and environmental quality as compatible goals, and to optimize environmental benefits. EQIP was originally authorized in the 1996 farm bill[1] as an amendment to the 1985 farm bill.[2] EQIP replaced four conservation programs repealed in the same law. These were the Great Plains Conservation Program, the Agricultural Conservation Program, the Water Quality Incentives Program, and the Colorado River Basin Salinity Control Program.

EQIP is the largest agriculture conservation program for working lands.[3] The program encourages farmers and ranchers to participate in conservation efforts by paying a portion of the cost of installing or constructing approved conservation practices. Eligible producers enter into EQIP contracts to receive payment for implementing conservation practices. Approved activities are carried out according to an EQIP plan developed in conjunction with the producer that identifies the appropriate conservation practice or practices to address resource concerns on the land.

EQIP was amended and reauthorized in both the 2002 and 2008 farm bills.[4] The U.S. Department of Agriculture's Natural Resources Conservation Service (NRCS) administers EQIP under a final rule.[5] NRCS implemented EQIP by establishing national priorities to reflect the most pressing natural resource needs and emphasize offsite benefits to the environment. The current national priorities, set in 2008 by NRCS, are as follows:

- reduction of nonpoint source pollutants in impaired watersheds (consistent with total maximum daily loads, or TMDLs);[6]
- conservation of ground and surface water resources;
- reduction of emissions that contribute to air quality impairment violations of National Ambient Air Quality Standards;
- reduction of soil erosion and sedimentation from unacceptable levels on agricultural land; and

- promotion of at-risk species habitat conservation.

Eligibility and Program Requirements

Producers with eligible land[7] can submit an EQIP plan that describes the conservation and environmental purposes that will be achieved using one or more USDA-approved conservation practices. USDA-approved conservation practices may involve structures, vegetation, or land management. Structural practices include the establishment, construction, or installation of measures designed for specific sites, such as animal waste management facilities, livestock water developments, and capping abandoned wells. Vegetative practices involve introduction or modification of plantings, such as filter strips or trees. Land management practices require site- specific management techniques and methods, such as nutrient management, irrigation water management, or grazing management.

Producers can receive technical assistance to develop an EQIP plan and, after approval, to implement the plan. Decisions about which plans to fund are made by USDA at the state level, with local input. Applications are accepted and ranked throughout the year within each state. Applications are grouped with similar crop, forestry, and livestock operation applications and evaluated within the groups. Additional funding groups may be created to rank applications based on similar resource objectives, geographic area, or type of agricultural operation. After an application is selected and approved, USDA provides payments to help the producer offset the cost of each practice, as well as income forgone relating to that practice implementation. Participants are eligible to receive payments for both constructing structures and implementing land management practices. Of the total annual EQIP spending, 60% is allocated to livestock practices.

Under an EQIP contract, USDA pays up to 75% of the projected costs associated with planning, design, materials, equipment, installation, labor, management, maintenance, or training, or up to 100% of the estimated income forgone to implement certain conservation practices. This payment rate can be higher for limited-resource, socially disadvantaged, or beginning farmers and ranchers,[8] provided this increase does not exceed 90% of practice costs. Initial payments are made in the year in which the contract is signed, but most payments are made after the practices are completed.

Contracts have a term of one to ten years and payments are limited by direct attribution to individuals or entities.[9] Total payments a person or entity can receive over any six-year period are limited to $300,000, except for projects having special environmental significance, which are limited to $450,000 over any six-year period. Individuals or entities with an average annual non-farm adjusted gross income (AGI) of $1 million or more for the three years prior to the contract period are ineligible unless they received at least two-thirds of their AGI from farming, ranching, or forestry. The 2008 farm bill created a case-by-case waiver to the AGI limitation for conservation programs (discussed in more detail in the "Adjusted Gross Income (AGI) Waiver" section below).

Program Funding

The 1996 farm bill authorized EQIP funding of $130 million in FY1996 and $200 million annually from FY1997 through FY2002. The 2002 farm bill significantly increased the annual authorized funding level incrementally from $400 million in FY2002 to $1.3 billion in FY2007. EQIP funding levels were revised in Section 1203 of the Deficit Reduction Act of 2005 (P.L. 109- 171) to limit funding to $1.27 billion in FY2007, while extending the authorization through FY20 10 and providing $1.27 billion in each of FY2008 and FY2009, and $1.3 billion in FY20 10.

The 2008 farm bill further increased the annual authorized funding levels incrementally from $1.34 billion in FY2009 to $1.75 billion in FY20 12. Funding under EQIP is mandatory (not subject to annual appropriations), and the program receives authorized amounts each year under the borrowing authority of USDA's Commodity Credit Corporation (CCC).[10] Congress, however, has limited EQIP funding below authorized levels in every year since FY2005, through annual appropriations bills. **Figure 1** identifies the authorized and actual funding levels for EQIP. The Senate-reported FY2009 agriculture appropriations bill (S. 3289, S.Rept. 110-426) would limit EQIP to $1.05 billion for FY2009—a reduction of $285 million from the authorized level of $1.34 billion in the 2008 farm bill.[11] Final action on FY2009 agriculture appropriations is pending.

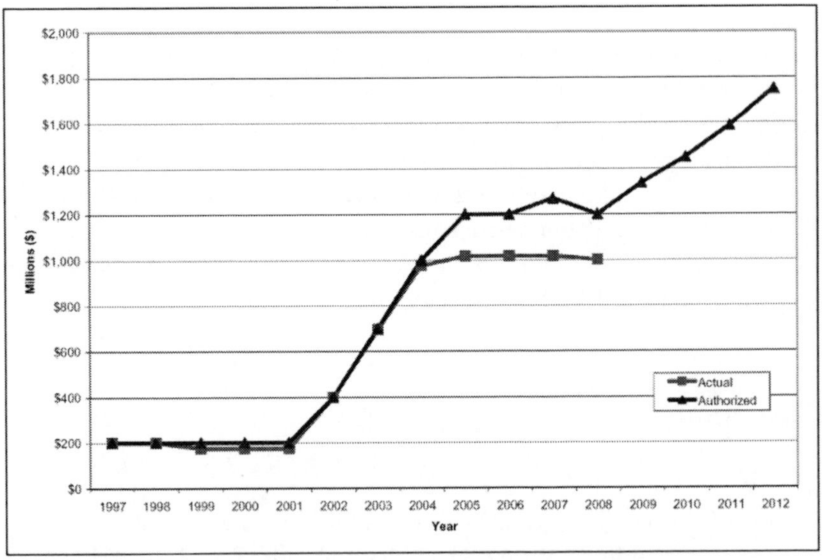

Source: CRS, based on historical Agriculture Appropriations Reports.
Note: Final action on FY2009 agriculture appropriations is pending.

Figure 1. EQIP Funding and Reductions, FY1997-FY2012($ in millions)

Table 1. Four Largest EQIP Allocation Recipient States, FY2003-FY2007 ($ in thousands)

Fiscal Year	Highest Allocation	2nd Highest Allocation	3rd Highest Allocation	4th Highest Allocation	Total Allocations
2003	Texas $57,717	California $48,582	Colorado $25,560	Nebraska $20,442	$626,701
2004	Texas $78,566	California $62,114	Colorado $36,932	Nebraska $29,600	$908,280
2005	Texas $90,007	California $62,114	Colorado $39,186	Minnesota $32,924	$991,879
2006	Texas $91,290	California $62,902	Colorado $41,200	Minnesota $32,000	$1,013,277
2007	Texas $89,124	California $62,090	Colorado $40,216	Minnesota $32,907	$1,004,926

Source: USDA, NRCS, *EQIP Program Information by Fiscal Year*, http://www.nrcs.usda.gov/programs/eqip/.

Notes: Allocations in this table represent financial assistance allocated to states only. Technical assistance, administrative, and technology costs are not included.

Annual funding received for EQIP is allocated to the states by NRCS using a formula based on national priorities, natural resource need, efficiency and performance measures, and regional equity.[12] The EQIP allocation formula uses 20 weighted factors based on the characteristics of agriculture and land use and resource considerations. Factors with the largest weights within the FY2009 formula include irrigated cropland, non-irrigated cropland, non-federal grazing land, livestock animal units, cropland eroding above the tolerance level, and impaired rivers and streams. States that receive the largest EQIP allocations have remained consistent from year to year, with Texas, California, and Colorado receiving the highest levels of funding annually between FY2003 and FY2007 (see **Table 1**).

Subprograms

Agricultural Water Enhancement Program

One of two subprograms under EQIP is the Agricultural Water Enhancement Program (AWEP). The 2008 farm bill (Sec. 2510, P.L. 110-246) created AWEP to promote ground and surface water conservation and to improve water quality on agricultural lands. The program replaces two previous water conservation programs: the Ground and Surface Water Conservation Program and the Klamath Basin Program.

Eligible partners or groups[13] submit project proposals to conserve ground and surface water or improve water quality in a specified area. NRCS selects projects based on requirements established in a *Federal Register* notice[14] and enters into agreements with selected partners. Once proposals for specific areas are selected, there are two methods for producers to sign up for an AWEP contract. Producers may either (1) apply directly to NRCS for approved agricultural water enhancement activities or (2) apply through the partner or group who submits applications on the producer's behalf. Funding is authorized as a separate amount from the general EQIP, at $73 million for each of FY2009 and FY2010, $74 million in FY2011, and $60 million in FY2012 and each fiscal year thereafter.

Conservation Innovation Grants

The second subprogram under EQIP is the Conservation Innovation Grants (CIG) program, which was created in the 2002 farm bill. The program is implemented through EQIP and is intended to leverage federal investment, stimulate innovative approaches to conservation, and accelerate technology transfer in environmental protection, agricultural production, and forest management. Examples of CIG projects include developing market-based approaches in conservation, demonstrating precision agriculture, capturing nutrients through a community anaerobic digester, and establishing a tribal partnership for regional habitat conservation.[15] The program was reauthorized in the 2008 farm bill through FY2012 at an unspecified funding level of general EQIP dollars. NRCS utilizes this discretion to determine the level of general EQIP funds used for CIG and allocates approximately $15 million for a national competition and up to $5 million for a Chesapeake Bay watershed competition annually. In addition, 32 states conduct, or have conducted, a state-level CIG competition, which has awarded over $17 million since FY2005. CIG awarded a total of $12.6 million (35 projects) in FY2004, $22.0 million (105 projects) in FY2005, $25.3 million (161 projects) in FY2006, $26 million (176 projects) in FY2007, and $21 million (95 projects) in FY2008. A recent announcement of program funding for CIG indicates that approximately $20 million is available in FY2009.[16]

The 2008 farm bill made some modifications to the CIG program. Previously, grants could not exceed 50% of the project cost, with nonfederal matching funds provided by the grantee. The 2008 farm bill removed this requirement, though USDA still requires a 50% match of nonfederal funds.[17] Also, the farm bill added an air quality component requiring that payments be made through CIG to producers to implement practices to address air quality concerns from agricultural operations and to meet federal, state, and local regulatory requirements. This air quality component is authorized at $37.5 million annually. The highest level of funding for all of CIG was in FY2006 ($26 million), making the $37.5 million requirement for air quality a potentially difficult target for the program.

SELECTED ISSUES

EQIP continues to receive widespread support in the farm community and in Congress, as it remains the major source of financial and technical assistance to help producers implement conservation practices that address specific resource and environmental problems. During the 111th Congress, several issues may attract congressional interest, including budgetary pressures, a continuing backlog of unfunded applications, adjusted gross income waivers, and measuring program accomplishments.

Mandatory Funding Levels

The 2008 farm bill reauthorized EQIP through September 30, 2012, with annual authorized funding levels of $1.2 billion in FY2008, $1.34 billion in FY2009, $1.45 billion in FY2010, $1.59 billion in FY2011, and $1.75 billion in FY2012. As shown in **Figure 1**, the authorized funding level has continued to increase since the 2002 farm bill; however, annual appropriations acts have reduced the actual funding levels by a total of nearly $820 million from FY2005 through FY2008.[18] With the 111th Congress facing tighter budget constraints, similar cuts to EQIP could be considered either in the appropriations process or if budget reconciliation is required.[19]

Unfunded Application Backlog

A main justification for the large funding increase in the 2002 farm bill was to respond to a large backlog of producer demand that had been documented during the farm bill debate. Despite this increase in funding, the number of pending applications continues to exceed the amount of available funding (see **Table 2**). Although this gap now constitutes a smaller portion of applications, it is still an issue for many producers who seek environmental assistance and are continuously denied funding due to budgetary constraints.[20] Many conservation groups worry that this could deter producers from applying and enrolling in the program. This issue will likely intensify if annual appropriations continue to reduce actual funding for EQIP.

Table 2. EQIP Funded and Unfunded Applications and Funds Obligated

Fiscal Year	Total Applications Funded	Total Applications Unfunded	Percentage of Applications Funded	Funds Obligated (Financial Assistance, $ in millions)
2000	16,249	37,712	30%	$139,606
2001	17,648	29,777	37%	$160,123
2002	19,817	70,495	22%	$322,193
2003	30,251	174,062	15%	$483,484
2004	46,413	135,394	26%	$718,150
2005	49,406	32,708	60%	$794,261
2006	41,190	32,633	56%	$787,968
2007	41,700	40,535	51%	$784,186

Source: USDA, NRCS, *EQIP Contract and Funding Information*, http://www.nrcs.usda.gov/programs/eqip/.

One reason why higher funding has not resulted in the elimination of the backlog is that the average contract size has grown since the 2002 farm bill. The average cost of an EQIP contract has more than doubled from almost $7,800 per contract prior to 2002 to over $16,000 per contract since 2002.[21] One reason for this increase could be the higher funding cap established in the 2002 farm bill that allowed large-scale livestock operations to fund waste management facilities and allowed the installation of more expensive conservation practices. According to NRCS, between 1997 and 2007, the top practice by cumulative cost-share dollars was waste storage facilities, which totaled $486 million over the ten-year period.[22] Though the 2008 farm bill lowered the payment limitation to $300,000 over any six-year period, the average contract is still considerably less ($16,000) than the limit. This will continue to be an issue as it is widely believed that the lower payment limitation will not greatly reduce the number of unfunded applications.

Adjusted Gross Income (AGI) Waiver

Another issue that the 111th Congress will likely monitor is the new waiver of the AGI limitation for conservation programs created in the 2008 farm bill. The AGI provision sets a maximum amount of income that an individual can earn and still remain eligible for program benefits, but USDA is allowed to waive the limit in certain cases. USDA administers the AGI limitation through an interim rule issued on December 29, 2008.[23]

Income limitations on conservation programs have been somewhat controversial. The 2008 farm bill made the AGI limitation for conservation programs higher than the AGI limitation for commodity programs to encourage environmental stewardship on farms and ranches, particularly larger operations that may have greater natural resource concerns. Supporters of AGI limits believe that tighter limits benefit small producers and gain additional public support for all agricultural programs through fiscal responsibility. Opponents of AGI limits on conservation programs believe that conservation benefits the general public, and thus any producer's enrollment, irrespective of wealth, is good for the general public.

The AGI waiver can be granted on a case-by-case basis by USDA if it is determined that environmentally sensitive land of special significance would be protected through the conservation program.[24] Under this rule, USDA can waive the AGI limitation through a written request that documents that the land contains critical resources, for example, endangered, threatened, or at-risk species; historical or cultural resources; unique wetlands; or critical groundwater recharge areas. The waiver must also show that the producer's participation is critical for the success of the project and will benefit multiple producers in the community, geographic region, or watershed. The number and frequency of AGI waivers granted is not limited, is at USDA's sole discretion, and remains to be determined.

Measuring EQIP Accomplishments

From available records, NRCS can provide considerable information about EQIP contracts, including which conservation practices are being installed, and their design and maintenance standards. However, until recently, relatively little was known about what is actually being accomplished through EQIP contracts. To begin filling this void, NRCS has compiled information about various resource concerns that EQIP addresses. These data show that in 2007, the primary resource concerns addressed through EQIP spending included water quality (20%), plant condition (17%), soil erosion (16%), water quantity (13%), domestic animals (12%), soil condition (10%), wildlife and fish (7%), and air quality (5%).[25] Little is known, however, about how enduring those conservation practices might be after the contract ends. Among the questions that NRCS is trying to address for all of its conservation activities, including EQIP, are how to (1) evaluate performance, (2) measure environmental changes, (3) evaluate cost-effectiveness, (4) determine which methods to use to identify environmental effects, and (5) determine which types of data should be collected to measure output.

NRCS initiated a national review in 2003, called the Conservation Effects Assessment Project (CEAP), in an attempt to develop better answers to all these questions. CEAP was originally intended to account for the benefits from the 2002 farm bill's substantial increase in conservation program funding through the scientific understanding of the effects of conservation practices at the watershed scale.[26] Only a few initial results are currently available; however, a draft watershed report is expected to be released this year for technical review.[27]

End Notes

[1] Section 334 of the Federal Agriculture Improvement and Reform Act of 1996, P.L. 104-127, 16 U.S.C. 3839aa.

[2] Sections 1240-1240I of the Food Security Act of 1985, P.L. 99-198.

[3] Working lands conservation programs are typically classified as programs that allow private land to remain in production, while implementing various conservation practices to address natural resource concerns specific to the area. Other conservation programs retire land from production or place restrictive easements on the land.

[4] Section 2301 of the Farm Security and Rural Investment Act of 2002, P.L. 107-171, and Section 2501 of the Food, Conservation, and Energy Act of 2008, P.L. 110-246.

[5] USDA, NRCS, "Environmental Quality Incentives Program," 74 *Federal Register* 2293, January 15, 2009.

[6] For more information on TMDLs, see CRS Report 97-831, *Clean Water Act and Total Maximum Daily Loads (TMDLs) of Pollutants*, by Claudia Copeland.

[7] Eligible land includes cropland, rangeland, pasture, non-industrial private forestland, and other land on which resource concerns related to agricultural production could be addressed through an EQIP contract.

[8] USDA combines these three groups and refers to them as "historically underserved producers." A limited resource producer or rancher is defined as having direct or indirect gross farm sales of less than $155,200 in each of the previous two years (adjusted for inflation) and a total household income at or below the national poverty level, or less than 50% of county median household income in the previous two years. A beginning farmer or rancher is defined as having farmed for less than 10 consecutive years. Socially disadvantaged farmers or ranchers are defined as having been subjected to racial or ethnic prejudice because of their identity as members of a group without regard to their individual qualities. Previously, USDA included gender prejudice in the definition of a socially disadvantaged farmer or rancher; however, changes in the 2008 farm bill removed gender from the definition, as it applies to conservation programs.

[9] Direct attribution means that payments must be directly attributed to a living person. If the person is part of a larger business entity then payments must be directly attributed to that person based on ownership shares in the entity. Individual people may receive EQIP payments through any number of contracts or ownership arrangement of farms, but the total amount of payments attributed to each living person may not exceed the statutory limits.

[10] The CCC is the funding mechanism for the mandatory payments that are administered by various agencies of USDA. For EQIP, NRCS provides the staff.

[11] For more information, see CRS Report R40000, *Agriculture and Related Agencies: FY2009 Appropriations*, coordinated by Jim Monke.

[12] The regional equity provision was first instituted in the 2002 farm bill (P.L. 107-17 1, Sec. 2701) and reauthorized in the 2008 farm bill (P.L. 110-246, Sec. 2703). The provision mandates that each state receive annually a minimum aggregate amount of funding for specified conservation programs. Regional equity affects not only EQIP but also the Wildlife Habitat Incentives Program, the Farmland Protection Program, and the Grassland Reserve Program. The 2008 farm bill increased the minimum level of funding to each state for these combined four conservation programs from $12 million to $15 million.

[13] An eligible partner or group may be a federally recognized tribe, state, unit of local government, agricultural or silvicultural association, or other such group of agricultural producers.

[14] USDA, NRCS, "Agricultural Water Enhancement Program," 74 *Federal Register* 2040, January 14, 2009.

[15] For additional examples of CIG projects, see http://www.nrcs.usda.gov/programs/cig/.

[16] USDA, NRCS, *Conservation Innovation Grants Fiscal Year (FY) 2009 Announcement for Program Funding*, Catalog of Federal Domestic Assistance (CFDA) Number: 10.912, January 16, 2009, http://www.nrcs.usda.gov/ programs/cig/pdf_files/Fiscal_Year_2009_ Announcement_for_Program_Funding.pdf.

[17] USDA, NRCS, *Conservation Innovation Grants Fiscal Year (FY) 2009 Announcement for Program Funding*, Catalog of Federal Domestic Assistance (CFDA) Number: 10.912, January 16, 2009, http://www.nrcs.usda.gov/ programs/cig/pdf_files/Fiscal_Year_2009 _Announcement_for_Program_Funding.pdf, p. 8.

[18] Annual appropriations reduce funding for several other agriculture mandatory programs as a means of meeting overall budget targets. The Senate-reported agriculture appropriations bill (S. 3289, S.Rept. 110-426) proposes an additional reduction of $285 million below authorized levels for FY2009.

[19] For more information on the budget reconciliation progress, see CRS Report RL30458, *The Budget Reconciliation Process: Timing of Legislative Action*, by Robert Keith.

[20] At the conclusion of FY2007 (most recent data available), states with the highest total of unfunded applications were Oklahoma (3,772), Texas (3,078), Missouri (2,813), Mississippi (2,163), and Montana (1,811).

[21] Contracts can vary from one to ten years; however, most are between two and five years in length and include between two and five practices. Data compiled by Soil and Water Conservation Society and Environmental Defense, *Environmental Quality Incentives Program (EQIP)—Program Assessment*, March 2007, http://www.swcs.org/ documents/filelibrary/EQIP_assessment.pdf.

[22] Other top practices between FY1997 and FY2007, by cumulative cost-share dollar, were irrigation systems ($337 million), fence ($329 million), brush management ($190 million), pipeline ($187 million), irrigation pipeline ($168 million), and nutrient management ($164 million). The term cost-share describes the percentage of the cost to install conservation practices paid by USDA. This term does not represent incentive payments and was removed from the program in the 2008 farm bill.

[23] USDA, CCC, "Farm Program Payment Limitation and Payment Eligibility for 2009 and Subsequent Crop, Program, or Fiscal Years," 73 *Federal Register* 79267-79284, December 29, 2008.

[24] Section 1604, P.L. 110-246.

[25] USDA, NRCS, *Environmental Quality Incentives Program: Program Information Review, Fiscal Year 2007.*

[26] Lisa F. Duriancik, Dale Bucks, and James P. Dobrowolski et al., "The First Five Years of the Conservation Effects Assessment Project," *Journal of Soil and Water Conservation*, vol. 63, no. 6 (Nov/Dec 2008), p. 185A.

[27] For more information on CEAP, see http://www.nrcs.usda.gov/technical/NRI/ceap/.

CHAPTER SOURCES

The following chapters have been previously published:

Chapter 1 – This is an edited, excerpted and augmented edition of a United States Congressional Research Service publication, Report Order Code RL34645, dated February 4, 2009.

Chapter 2 – This is an edited, excerpted and augmented edition of a United States Congressional Research Service publication, Report Order Code R40202, dated February 6, 2009.

Chapter 3 – This is an edited, excerpted and augmented edition of a United States Congressional Research Service publication, Report Order Code R40187, dated February 3, 2009.

Chapter 4 – This is an edited, excerpted and augmented edition of a United States Congressional Research Service publication, Report Order Code R40186, dated February 3, 2009.

Chapter 5 – This is an edited, excerpted and augmented edition of a United States Congressional Research Service publication, Report Order Code R40168, dated February 13, 2009.

Chapter 6 – This is an edited, excerpted and augmented edition of a United States Congressional Research Service publication, Report Order Code R40242, dated February 23, 2009.

Chapter 7 – This is an edited, excerpted and augmented edition of a United States Congressional Research Service publication, Report Order Code R40197, dated February 5, 2009.

INDEX

A

AAA, 79
abatement, 121, 122, 126, 127, 129, 130, 131, 133, 138, 143, 152, 154, 156, 158, 168, 170
accidents, 28
adaptation, xii, 82
adjusted gross income, 182, 185
adjustment, 166
administration, 42
administrative, xiv, 6, 18, 25, 35, 41, 118, 135, 136, 137, 145, 148, 149, 153, 166, 183
AEI, 172, 177
Africa, 90
AGI, 182, 187
agricultural, xii, xiv, 82, 85, 86, 87, 89, 90, 91, 97, 101, 102, 107, 150, 151, 152, 155, 161, 179, 180, 181, 184, 185, 187, 189
agricultural sector, 87, 91
agriculture, xiv, 82, 85, 86, 91, 114, 152, 179, 180, 182, 183, 184, 185, 190
aid, vii, x, 49, 164
air, xii, 10, 31, 57, 66, 86, 93, 95, 98, 101, 121, 129, 152, 153, 180, 185, 188
air pollutant, 66, 121, 129
air pollution, 152
air quality, xii, 57, 86, 93, 95, 98, 101, 180, 185, 188
Alaska, 58
Albania, 56
alternative, ix, x, xii, xiii, 16, 18, 25, 26, 27, 28, 29, 30, 32, 35, 36, 38, 44, 82, 87, 88, 93, 94, 95, 96, 97, 98, 99, 100, 102, 104, 105, 106, 109, 110, 112, 114, 124, 141, 149, 156
alternative energy, xii, 82, 87, 88
alternatives, 18, 19, 28, 37, 38, 43, 90, 94, 95, 99, 136, 165
aluminum, 177
Amazon, 130
amendments, 32, 97, 99
American Jobs Creation Act, 99, 107
AMT, 107, 110, 111
analysts, 12, 155
animals, 152, 188
anthropogenic, 155
API, 65, 78, 79
application, ix, xi, 15, 16, 17, 18, 20, 21, 22, 23, 41, 81, 85, 87, 88, 133, 155, 164, 176, 181
appropriations, xiv, 6, 20, 21, 33, 88, 108, 179, 182, 183, 186, 190
appropriations bills, 33, 182
Appropriations Committee, 33, 40
Areva, 29, 36, 37
argument, 126, 129, 133, 135, 138, 139, 142, 143, 149, 164, 165, 167

Army Corps of Engineers, 3
ARS, 83, 89, 91
assessment, 27, 90, 190
assumptions, xiv, 118, 126, 129, 139, 155, 156, 158, 166
Atlantic, 40, 47, 68, 86
atmosphere, xi, 38, 81, 84, 90, 91, 129, 130, 138, 145, 151, 153, 155, 156, 161
Atomic Energy Act, 5, 33, 34
Atomic Energy Commission, 3, 4, 7, 8, 11, 38, 41, 47
attribution, 181, 189
authority, 18, 26, 27, 31, 32, 33, 34, 105, 112, 121, 124, 125, 137, 139, 182
automobiles, 59, 99, 149
availability, xii, 66, 81, 85, 86, 103
aviation, 59, 60

B

back, 22, 36, 61
banking, 64, 123, 132, 133, 171
Barack Obama, 17, 44
barrier, 103, 112
barriers, xii, 93, 107, 108
batteries, 95, 98
battery, 95, 97, 106, 109, 110
behavior, 42, 135, 159, 160
behavioral change, 120, 169
benefits, xi, xii, xiv, 19, 36, 41, 81, 84, 87, 93, 108, 112, 118, 121, 122, 126, 127, 129, 130, 131, 133, 148, 154, 156, 160, 165, 166, 170, 175, 176, 180, 187, 188
biodiesel, 97, 98, 99, 107, 110, 113, 114, 115
biofuels, xiii, 94, 95, 96, 97, 99, 102, 103, 111, 115
biomass, xiii, 84, 86, 89, 94, 96, 100, 102, 103, 107
biorefinery, 98
blends, 107
bomb, vii, viii, 1, 2, 3, 9, 11
boreholes, 38
borrowing, 132, 182
Brazil, 103
breakdown, 28, 59, 83
broad spectrum, 75, 145
Btus, 51, 73, 77
Budget Committee, 140
buildings, 120, 153
Bulgaria, 56
Bureau of Economic Analysis, 5
Bureau of Land Management, 34, 46
burn, 57, 114
Bush Administration, 20, 22, 108, 115
butyl ether, 73, 100

C

Canada, 55
CAP, 173
capital cost, xi, 50, 67
caps, 19, 69, 123, 130, 172
Carbon, v, vii, xiv, 44, 84, 87, 89, 90, 91, 117, 118, 125, 134, 135, 139, 144, 150, 154, 158, 159, 163, 164, 169, 170, 171, 172, 173, 174, 175, 176, 177
carbon credits, 87
carbon dioxide, 26, 85, 87, 115, 170
carbon emissions, 87, 136
carbon monoxide, 115
Caribbean, 103, 115
Caribbean Basin Initiative, 103, 115
Caribbean Basin Initiative (CBI), 103, 115
CCC, 182, 189, 190
cell, 98, 105, 106, 108, 109, 114, 115
cellulose, 102
cellulosic, 96, 97, 99, 102, 103, 115
cellulosic ethanol, 103
cement, 150, 152, 153, 174, 175
CEO, 173
CH4, 85, 150, 170
charcoal, xi, 81, 82, 83
Chile, 90
China, 63, 142, 143
civilian, 3, 5, 12
classes, 112, 153
Clean Air Act, 98, 100, 101, 121, 174
Clean Development Mechanism, 173
Clean Water Act, 189

cleanup, 43, 102
climate change, xi, xii, xiv, 9, 10, 81, 82, 87, 88, 94, 95, 118, 119, 120, 127, 129, 130, 133, 138, 142, 154, 155, 156, 160, 166, 171, 175
Co, 37, 45, 47, 79, 99, 142
CO2, 85, 87, 118, 119, 126, 129, 130, 133, 135, 138, 143, 144, 145, 147, 148, 149, 150, 155, 158, 163, 168, 169, 170, 171, 172, 173, 174, 175, 176, 177
coal, x, xi, 50, 51, 54, 57, 59, 66, 68, 69, 72, 114, 115, 119, 120, 146, 147, 149, 150, 151, 152, 153, 158, 164, 169, 174
coal mine, 150, 151, 152, 153
Cold War, 9
Colorado, 33, 46, 180, 183, 184
Columbia, 45, 100
combustion, 100, 114, 119, 135, 145, 149, 150, 159
commercialization, xii, 10, 93, 108
Committee on Appropriations, 47, 111
Committee on Environment and Public Works, 140
commodities, 61, 64
commodity, 60, 104, 152, 187
Commodity Credit Corporation, 182
Commodity Credit Corporation (CCC), 182
communities, 86, 88, 161
community, 86, 185, 188
compensation, 30
competition, 165, 185
complement, 142
complexity, 135, 155
compliance, 105, 115, 123, 124, 141, 167, 172
complications, 80
components, 18, 95, 98, 132, 137
composition, 75
compounds, 115
concentration, 42, 129, 130, 138, 155, 175
conceptualization, 140
concrete, viii, 2, 31
conditioning, 153
Conference Committee, 98
confidence, x, 16, 26, 29
Congressional Budget Office, 46, 140, 157
Congressional Record, 47
Connecticut, 26
consensus, 9, 24
conservation, xii, xiv, 82, 88, 89, 98, 142, 148, 161, 169, 179, 180, 181, 182, 184, 185, 186, 187, 188, 189, 190
constraints, xiv, 95, 145, 179, 186
construction, xi, 20, 22, 26, 31, 50, 67, 111, 181
consumers, xiii, 57, 69, 70, 117, 119, 120, 133, 137, 141, 146, 163, 173, 174
consumption, vii, x, xi, xiii, 9, 49, 50, 51, 52, 56, 57, 59, 60, 64, 65, 66, 69, 73, 75, 77, 80, 94, 95, 96, 102, 112, 114, 133, 145, 146, 147, 169
contamination, 101
contracts, 20, 25, 45, 180, 188, 189
control, xiii, xiv, 9, 12, 21, 117, 118, 119, 120, 121, 123, 124, 125, 126, 131, 133, 135, 138, 139, 140, 141, 143, 144, 145, 148, 152, 154, 160, 166, 167, 170, 171, 172
conversion, 106, 110, 111, 169
cooking, 84, 85, 107
cooling, 30, 31
corn, xiii, 86, 94, 96, 101, 102, 103
correlation, 131
corruption, 144
cost analyses, 158
cost curve, 122, 126, 127, 130, 131, 170
cost-effective, 86, 90, 188
counterbalance, 95
Court of Appeals, 45
covering, 145, 148
credit, xiii, 88, 89, 94, 95, 99, 106, 107, 110, 111, 114
criticism, 17, 40
crop residues, xi, 81, 82, 83, 86
crops, 102
crude oil, 61, 63, 64, 79, 94, 99, 151, 169
crystalline, 40
currency, 170
current limit, ix, 16
customers, 146

Customs Service, 115
Czech Republic, 56

D

debates, ix, 3, 35
decay, 18, 28, 36, 131
decisions, xiv, 29, 34, 35, 40, 42, 45, 118, 131, 137, 144, 155, 166, 173, 174
defense, ix, 16, 19, 20, 24, 25, 27, 30, 37, 42, 43, 119
Defense Authorization Act, 105
Deficit Reduction Act, 182
definition, 38, 105, 129, 130, 155, 189
delivery, 63
Department of Agriculture, xii, xiv, 83, 88, 89, 91, 94, 98, 114, 151, 152, 179, 180
Department of Energy, ix, 6, 11, 12, 15, 17, 42, 44, 45, 46, 47, 78, 102, 105, 113, 115
Department of Energy (DOE), ix, 6, 15, 17, 105
Department of the Interior, x, 16, 34
Department of Transportation, xii, 94
developed countries, 88
developed nations, 144
developing countries, 85, 88
developing nations, 143, 144, 166
dichotomy, 123
diesel, 59, 60, 65, 72, 79, 95, 97, 98, 99, 100, 103, 104, 105, 106, 107, 109, 112, 115
diesel engines, 95, 97, 107
diesel fuel, 59, 65, 79, 98, 100, 103, 104, 107
direct cost, 25
discharges, 19
discount rate, 155, 175
discounting, 131
distortions, 159, 160
distribution, 133, 139, 161, 162, 164, 165, 176, 177
distribution of income, 133
District of Columbia, 45, 100
District of Columbia Circuit, 45
diversity, viii, 2, 9, 11
dominance, viii, 2, 11
draft, 38, 188
dumping, 23, 38

E

earth, 38, 90, 155, 161
Eastern Europe, 56
economic activity, 75
economic crisis, 64
economic efficiency, xiv, 118, 121, 133, 142, 143, 145, 160, 162, 166, 177
economic goods, 119
economic growth, 75, 77, 122
economic incentives, 10, 99
economic problem, 77
Economic Research Service, 114
economic theory, 121, 122, 154, 155
economics, xii, 93, 119, 126, 157
EEG, 30
election, ix, 15, 42
electric energy, 80
electric power, xi, 50, 56, 59, 66, 70, 72, 73, 78, 169
electric utilities, 26, 70, 160
Electric utilities, 146
electricity, xi, 50, 51, 52, 65, 66, 67, 68, 69, 70, 72, 73, 74, 75, 80, 108, 133, 146, 150, 162, 164, 166, 174
electromagnetic, 38
embargo, 4, 61
emission source, 91, 120, 124, 127, 136, 145, 147, 148, 152, 174
emitters, 87
employment, 119, 159
end-users, 169
Energy and Water Development, 31, 33, 47
energy consumption, xi, 9, 50, 51, 52, 65, 75, 80, 133, 147
energy efficiency, xi, 50, 51, 88, 119, 160, 166
Energy Efficiency and Renewable Energy, 115
Energy Independence and Security Act, xiii, 94, 96, 97, 102, 103, 108, 114

Energy Information Administration, 5, 12, 52, 53, 78, 80, 172
Energy Information Administration (EIA), 52, 53, 78, 80
energy markets, vii, x, 49, 50, 133
Energy Policy Act, 96, 97, 100, 101, 102, 104, 105, 106, 107, 108, 112, 113, 114, 115
Energy Policy Act of 2005, 96, 97, 100, 101, 102, 105, 106, 107, 108, 112, 114, 115
Energy Policy and Conservation Act, 98
engines, 72, 95, 97, 107
England, 36, 37, 68
environment, 4, 17, 24, 27, 28, 31, 38, 147, 180
environmental effects, 188
environmental impact, viii, 2, 11, 27, 37, 98, 99, 102
environmental issues, 6, 96
environmental policy, 140
environmental protection, viii, 2, 10, 11, 184
Environmental Protection Agency, xii, 23, 44, 45, 75, 91, 94, 100, 114, 121, 125, 174
Environmental Quality Incentives Program, v, vii, xiv, 179, 180, 189, 190
environmental standards, 41
environmentalists, 107
enzymatic, 102
EPA, 23, 41, 45, 75, 100, 109, 114, 121, 137, 145, 146, 150, 151, 152, 157, 174, 175
EQIP, v, vii, xiv, 179, 180, 181, 182, 183, 184, 185, 186, 187, 188, 189, 190
equity, 184, 189
estimating, 155, 158, 175
ethanol, xi, 50, 51, 73, 95, 96, 97, 98, 99, 100, 101, 103, 104, 110, 114, 115
Ethanol, 74, 100, 102, 103, 113, 114, 115
ethers, 100
EU, 91, 133, 141, 142, 143, 172, 173
Eurasia, 56
Europe, 56, 63, 176
European Union, 133, 134, 141
excise tax, 99, 107
exercise, 41, 125
explosives, 18
external costs, 119
extraction, 147, 174

F

failure, 31, 41, 119
Farm Bill, xii, xiii, 87, 94, 95, 97, 111, 114
Farm Security and Rural Investment Act, 189
farmers, xiv, 107, 179, 180, 181, 189
farming, 86, 152, 182
farmland, 102
Farmland Protection Program, 189
farms, 187, 189
federal budget, 24
federal funds, 165
federal government, ix, 16, 20, 25, 43, 44, 82, 108, 125, 177
federal grants, xii, 93
federal law, 34
Federal Register, 45, 115, 184, 189, 190
fee, xiii, 117, 118, 140, 145, 170
feedstock, xii, 81, 83, 84, 85, 86, 100, 147
fees, 20, 25, 139, 140
fermentation, 85, 102, 150
fertility, 83, 84, 85, 88
fertilizer, xi, xii, 81, 82, 84, 86, 87, 89, 152
fertilizers, 102, 152
financial crisis, 64
financial loss, 164
fire, 31, 57, 59, 146, 151, 160, 164, 174
firms, 133
fission, 35, 36
flight, 9, 12
flow, 43, 119, 129, 130, 138, 143
fluctuations, xiv, 70, 118, 125, 134, 138, 167
focusing, 10, 20, 40
Ford, 108, 112
forest management, 185
forestry, xii, 82, 86, 91, 161, 180, 181, 182

fossil, xiii, 74, 99, 117, 118, 119, 135, 136, 145, 147, 149, 150, 154, 158, 160, 164, 166, 169, 174, 175
fossil fuels, xiii, 99, 117, 118, 119, 136, 145, 147, 149, 154, 158, 166, 169, 175
framing, 140
France, 35, 37
Franklin D. Roosevelt, 3
fuel cell, 98, 105, 106, 108, 109, 114, 115
fuel cycle, 18, 35
fuel management, 20
full capacity, 165
funding, viii, ix, xiv, 2, 4, 6, 7, 8, 10, 11, 12, 16, 17, 18, 20, 22, 29, 30, 32, 95, 102, 108, 109, 115, 135, 160, 161, 165, 179, 181, 182, 183, 185, 186, 187, 188, 189, 190
funds, 20, 22, 99, 102, 185

G

gas, x, xi, xiii, 26, 33, 36, 49, 50, 51, 52, 54, 57, 66, 68, 69, 70, 77, 79, 81, 82, 84, 85, 87, 88, 91, 94, 96, 100, 102, 103, 107, 110, 114, 117, 118, 120, 129, 147, 150, 151, 153, 158, 166, 169
gas-cooled, 33, 36
gases, 68, 129, 148, 170
gasoline, x, 49, 51, 54, 59, 60, 61, 62, 63, 64, 65, 73, 79, 94, 95, 96, 97, 98, 99, 100, 101, 103, 104, 106, 112, 114, 115, 119, 154, 169
GDP, viii, 2, 7, 8, 11, 61, 62, 72, 75, 76, 77, 157, 161
General Motors, 108, 112
General Services Administration, 105
generation, xi, 50, 51, 57, 65, 66, 67, 68, 70, 73, 74, 78, 88, 134, 150, 164
generators, 66, 70
Georgia, 89
Germany, 3, 9
GHG, xiii, 87, 117, 118, 119, 120, 121, 125, 126, 127, 129, 130, 133, 135, 136, 137, 138, 139, 140, 141, 142, 143, 144, 148, 149, 150, 152, 153, 154, 156, 160, 165, 166, 167, 170, 171, 174, 175, 176
global climate change, 83
Global Nuclear Energy Partnership, 29, 36, 46
Global Nuclear Energy Partnership (GNEP), 36
global warming, 45, 85, 132, 170
Global Warming, 171, 172, 174, 176
GNEP, 36, 37, 46
goal setting, viii, 2
goals, viii, ix, 2, 3, 4, 9, 11, 40, 105, 138, 148, 166, 180
goods and services, 4, 155
government, iv, ix, 10, 16, 20, 25, 28, 29, 38, 43, 44, 61, 75, 82, 96, 103, 105, 108, 119, 120, 125, 137, 165, 170, 177, 189
Government Accountability Office, 113
governors, 101
GPO, 45, 46, 47
grants, xii, 31, 32, 87, 93, 95, 97, 102, 105, 109, 185
grasses, 96, 97, 102
grazing, 85, 181, 184
greenhouse, xi, xii, xiii, 81, 82, 84, 85, 87, 88, 91, 93, 94, 96, 103, 107, 117, 118, 166
Greenhouse, v, vii, 85, 91, 117, 150, 151, 152, 159, 169, 170, 171, 172, 173, 174, 177
greenhouse gas, xi, xii, xiii, 81, 82, 84, 85, 87, 88, 91, 93, 94, 96, 103, 107, 117, 118, 166
greenhouse gas (GHG), xiii, 87, 117, 166
greenhouse gases, xii, 85, 93
Greenland, 130
gross domestic product, viii, 2, 8, 11, 61, 161
Gross Domestic Product, 5, 7
Ground and Surface Water Conservation Program, 184
ground water, 85, 101
groups, 28, 101, 141, 142, 167, 181, 186, 189
growth, 40, 51, 59, 75, 77, 96, 98, 122

GSA, 109
GSM, 174
Gulf Coast, 39, 99

H

habitat, 181, 185
handling, xii, 17, 81, 85, 86
hanging, 171
harvest, 89
harvesting, 102
hazards, 44, 82
health, 28, 86, 155
heat, 36, 74, 108
heating, 51, 61, 79, 84
heating oil, 61, 79
heavy oil, 55
heavy water, 36
high temperature, xi, 81, 82, 83
high-level, ix, 15, 16, 18, 19, 26, 27, 37, 38, 39, 42, 43
historical trends, x, 49, 54
House, xii, 32, 33, 42, 46, 47, 93, 95, 97, 109, 110, 111, 140, 171, 173
House Appropriations Committee, 33
household, 90, 162, 163, 166, 176, 189
household income, 162, 163, 176, 189
households, xiii, 90, 117, 119, 120, 148, 160, 161, 162, 164, 166, 174, 177
housing, 64
human, 4, 28, 86, 102, 142, 155
humans, 38, 101
Hungary, 56
Hurricane Katrina, 63
hybrid, 96, 98, 105, 106, 108, 112, 114, 115, 147
hybrids, 98, 99, 105, 106, 112, 114
hydro, 68
hydroelectric power, 73
hydrogen, 98, 108, 110, 114, 115
hydropower, xi, 50, 51, 67, 68

I

id, 4, 33, 41, 113, 115, 133
Idaho, 27, 33
IEA, 56
implementation, xii, xiv, 42, 85, 94, 118, 120, 126, 135, 143, 144, 148, 166, 167, 175, 181
imports, xi, 50, 51, 63, 94, 98, 103, 104, 108, 115, 151, 174
incentive, 103, 115, 119, 149, 165, 190
incentives, xii, xiii, 93, 94, 98, 103, 106, 112, 119, 161
inclusion, 133, 148
income, xiv, 119, 133, 159, 160, 161, 162, 164, 176, 177, 179, 181, 182, 185, 187, 189
income tax, 119, 159, 160, 176
incomes, 97, 163
independence, 9
India, 142, 143
Indian, 32, 34, 46
Indians, 34, 46
indicators, 50
industrial, xiv, 51, 52, 70, 145, 147, 150, 164, 179, 189
industrial sectors, 147, 164
industry, x, xiii, 6, 25, 29, 33, 49, 51, 52, 61, 66, 74, 78, 79, 86, 101, 102, 113, 114, 118, 119, 135, 141, 146, 164, 166, 173, 174, 176
inefficiency, 127, 129, 132
infancy, xii, 81, 89
inflation, 6, 61, 77, 133, 189
infrastructure, 95, 98, 99, 104, 110, 112
initiation, 137
injection, 38
Innovation, 89, 184, 190
insight, 88
institutions, 37, 82, 88, 136
instruments, 126, 167
interaction, 176
interaction effect, 176
Intergovernmental Panel on Climate Change, 151, 154, 155, 170, 175, 176

Intergovernmental Panel on Climate Change (IPCC), 154, 155
internal combustion, 114
Internal Revenue Service, 100, 113, 114, 115, 176
International Energy Agency, 79
international trade, 79
inventories, 19, 36
investment, vii, viii, xiv, 1, 2, 9, 89, 118, 119, 133, 141, 159, 161, 162, 165, 166, 184
IPCC, 154, 155, 175
IPPC, 156
Iran, 61
iron, 150, 152, 153
irradiation, 35
irrigation, 181, 190
isolation, 17, 24, 42, 162
isotope, 35
isotopes, 18, 35, 39

J

Japan, 3
Japanese, 90
Jefferson, 113
jet fuel, 59, 104
jurisdiction, 26, 140, 167
jurisdictions, 167
justification, 23, 186

K

Katrina, 63
Kentucky, 26
Kenya, 90
kerosene, 59
Kyoto Protocol, 142, 173, 174

L

labor, 119, 159, 161, 162, 176, 177, 181
labor force, 159
land, xiv, 4, 9, 43, 85, 87, 179, 180, 181, 184, 187, 189
land disposal, 43
land use, 87, 184
landfills, 150, 152, 153
language, 33, 100, 111
large-scale, 6, 187
law, x, xiii, 16, 18, 21, 28, 31, 32, 34, 41, 43, 94, 96, 101, 103, 104, 105, 107, 180
laws, 26, 27, 98, 153
lawsuits, 41, 121
leakage, 149, 152, 167, 172
legislation, ix, x, 16, 18, 24, 25, 28, 32, 41, 87, 97, 101, 115, 120, 137, 140, 158, 167
legislative proposals, 17, 114, 158, 165
licenses, 34
licensing, ix, 16, 18, 22, 23, 26, 28, 34, 43
lifecycle, xiii, 94, 96, 103
light trucks, 75, 98, 120
limitation, 182, 187
limitations, xiv, 103, 179, 187
liquefied natural gas, 151
litigation, 40, 137
livestock, 148, 150, 152, 181, 184, 187
loan guarantees, 102
loans, xii, 93, 102
local government, 105, 189
location, 17, 18, 39, 115, 145, 149, 155
London, 38
losses, 127, 129, 154, 162, 164
Louisiana, 68
lower-income, 162
LPG, 99

M

macroeconomic, 161, 162
magnetic, iv
maintenance, xii, 28, 32, 33, 44, 81, 153, 181, 188
management, ix, 16, 20, 24, 27, 28, 29, 34, 38, 41, 85, 88, 150, 180, 181, 185, 187, 190
management practices, 181
mandates, xiii, 94, 96, 189

Manhattan, v, vii, viii, 1, 2, 3, 4, 6, 7, 8, 9, 10, 11, 12
manipulation, 68, 137
manufacturer, 104, 106, 110
manufacturing, 104, 109, 150, 152, 153, 160
manure, xi, 81, 82, 83, 85, 86, 88, 150, 152
marginal cost curve, 126, 127, 130, 131, 156, 170, 171
marginal costs, 121, 122, 127, 129, 138, 154, 158
market, x, xiii, 49, 55, 60, 64, 68, 75, 79, 95, 99, 101, 103, 117, 119, 120, 121, 122, 124, 130, 133, 136, 137, 142, 159, 160, 166, 172, 174, 185
market prices, 119, 134
marketplace, 6, 10, 165
markets, vii, x, 49, 50, 89, 124, 133
Massachusetts, 121, 132
Massachusetts Institute of Technology, 132
matching funds, 185
measurement, 145, 150, 152
measures, 105, 124, 131, 161, 181, 184
meta-analysis, 155, 156
methane, 85, 151, 152, 153, 170
methanol, 96, 113
methyl tertiary, 73, 100
metric, 19, 20, 25, 27, 31, 33, 34, 37, 39, 91, 147, 155, 161, 170, 175
metropolitan area, 115
Mexico, 3, 30, 43, 48, 55
Middle East, 55, 63
military, 5, 9
million barrels per day, 51, 59, 80
mines, 150, 151, 152, 153
Minnesota, 183
Mississippi, 190
Missouri, 190
MIT, 46, 132, 157, 159, 171, 174
mitigation policy, 120
models, 23, 112, 138, 176
modules, 12
momentum, xiv, 118, 139, 143
money, 41, 170
Montana, 190
motor system, 112
movement, 33, 50
MRS, 20, 25, 31, 32, 40, 41
MTBE, 73, 100, 101, 114, 115

N

NAS, 24, 30
NASA, 4, 7, 8, 12
nation, vii, ix, 1, 3, 4, 6, 15, 17, 39, 44, 100, 101, 144
National Academy of Sciences, 13, 24, 30, 35, 46, 171
National Academy of Sciences (NAS), 24, 30
National Aeronautics and Space Administration, 12
National Ambient Air Quality Standards, 180
National Defense Authorization Act, 105
National Income and Product Accounts, 12
national interests, 4
national policy, 35
National Research Council, 46
natural, xi, xii, 26, 31, 43, 50, 51, 54, 57, 63, 66, 68, 69, 70, 77, 79, 82, 96, 110, 114, 120, 147, 150, 151, 152, 153, 158, 169, 180, 183, 187, 189
natural disasters, 63
natural gas, xi, 26, 50, 51, 54, 57, 66, 68, 69, 70, 77, 79, 96, 110, 114, 120, 147, 150, 151, 153, 158, 169
Natural Resources Conservation Service, xiv, 89, 179, 180
NEA, 38
Nebraska, 183
NEMS, 157
net present value, 155, 156
neutrons, 36
Nevada, ix, 15, 17, 21, 22, 23, 31, 32, 41, 45, 47
New England, 68
New Jersey, 26, 89, 114
New Mexico, 3, 30, 43, 48
New World, 7, 8, 11
New York, iii, iv, 12, 44, 63, 101

New York Mercantile Exchange, 63
nitrogen, 129, 170
nitrous oxide, 85, 152, 170
nonproliferation, 36, 44
Norway, 153
NRC, ix, 5, 15, 16, 17, 19, 20, 21, 22, 23, 25, 26, 27, 29, 30, 31, 34, 35, 41, 43, 78
nuclear, viii, ix, x, xi, 2, 3, 4, 9, 11, 15, 16, 17, 18, 19, 20, 21, 23, 24, 25, 26, 27, 28, 29, 30, 31, 32, 33, 34, 35, 36, 37, 38, 39, 40, 41, 42, 43, 44, 45, 50, 51, 54, 55, 67, 78
Nuclear Energy Agency, 37, 47
Nuclear Energy Agency (NEA), 37
nuclear power, ix, xi, 4, 15, 16, 18, 19, 20, 24, 26, 32, 33, 40, 50, 51, 67, 78
nuclear power plant, ix, 15, 16, 18, 19, 20, 24, 26, 33
nuclear program, 27
nuclear reactor, 21, 30, 31, 35, 78
Nuclear Regulatory Commission, ix, 5, 15, 17, 21, 45, 46, 47, 67, 78
Nuclear Waste Policy Act, ix, 15, 18, 46
nuclear weapons, x, 3, 9, 16, 18, 27, 33, 35, 36, 44
nutrient, 181, 190
nutrients, 85, 185
NWPA, ix, 15, 16, 18, 19, 20, 21, 22, 23, 24, 25, 28, 31, 32, 34, 37, 39, 40, 41, 42
NYMEX, 63

O

obligations, 29
octane, 100
ODS, 153
OECD, 79
Office of Management and Budget, 7, 140
Office of Technology Assessment, 47
oil, vii, viii, x, xi, xii, 1, 2, 4, 5, 6, 11, 49, 50, 51, 52, 54, 55, 56, 60, 61, 63, 64, 66, 67, 68, 69, 75, 77, 79, 83, 85, 86, 89, 93, 94, 95, 96, 98, 99, 101, 105, 107, 112, 120, 147, 151, 169, 174
Oil and Gas Journal, 56
oil sands, 55
Oklahoma, 63, 68, 190
OPEC, 60
opposition, xi, 18, 26, 39, 40, 44, 50, 67, 141, 148
organic, xi, 81, 82, 83, 97, 107
organic matter, 83, 97
Organization for Economic Cooperation and Development, 37, 47
oversight, xii, 29, 46, 94, 124, 140
ownership, 29, 31, 189
oxide, 85, 129, 152, 170
oxygen, 83, 100, 101
ozone, 100, 114, 150, 153
Ozone, 151, 175

P

Pacific, 39, 45, 68
parameter, 132
Paris, 47
partnership, 108, 185
partnerships, 118
passenger, 75, 96, 98, 104, 106, 112, 115
pasture, xiv, 179, 189
payroll, 119, 160, 164, 177
peer, 156, 175
Pennsylvania, 11
perception, viii, 2, 6, 55
perceptions, x, 49, 167
periodic, 125, 138, 166
pesticides, 85
petrochemical, 57
petroleum, x, xiii, 5, 49, 54, 55, 56, 57, 59, 60, 63, 68, 78, 94, 95, 96, 98, 99, 100, 101, 102, 103, 104, 108, 110, 114, 148, 151, 158
Petroleum, 54, 56, 57, 58, 59, 60, 63, 65, 66, 68, 78, 79, 113, 150
petroleum products, 60
PFC, 170
pipelines, 151
planning, 26, 133, 181
plants, ix, xi, 15, 16, 18, 19, 20, 24, 26, 34, 36, 37, 43, 50, 51, 57, 66, 67, 74, 100,

103, 111, 119, 121, 141, 146, 147, 149, 150, 151, 161, 164, 174
play, xii, 24, 87, 94, 133, 140, 167
plug-in, xiii, 94, 95, 98, 105, 106, 110, 112
plutonium, x, 3, 16, 18, 27, 35, 36, 42, 44
Poland, 56
polar ice, 38
policy instruments, 120
policy makers, 75
policy making, vii, x, 49
political subdivision, 144
politics, 29
pollutant, 129, 130, 138, 168, 170
pollutants, 180
pollution, 100, 119, 152, 154, 159
pools, 19, 30
poor, 95, 130
population, 100, 122
poverty, 189
power, ix, xi, 4, 15, 16, 18, 19, 20, 24, 26, 32, 33, 40, 50, 51, 56, 59, 65, 66, 67, 68, 70, 72, 73, 78, 106, 110, 119, 121, 141, 145, 147, 149, 150, 169
power generation, 57, 65, 66, 68, 73
power plant, ix, 15, 16, 18, 19, 20, 24, 26, 33, 57, 119, 121, 141, 145, 147, 149, 150
Power plant, 150, 174
power plants, ix, 15, 16, 18, 19, 24, 26, 57, 119, 121, 141, 145, 147, 149, 150
PPA, 157
pre-existing, 144
preference, xiii, 104, 117, 123, 131, 141, 166
prejudice, 189
present value, 155, 156
President Bush, xiii, 94, 95, 96, 97, 103, 104, 105, 108
President Clinton, 32, 147
presidential veto, 33
press, 45, 46
pressure, 18, 34, 40, 63, 97, 142
price caps, 69
price ceiling, xiii, 118
price deflator, 157
price effect, 169, 173
price floor, 124
price index, 11
price stability, 126
prices, vii, viii, x, xiii, 2, 4, 6, 11, 26, 49, 55, 56, 60, 61, 63, 64, 66, 68, 69, 70, 78, 79, 94, 95, 96, 97, 98, 99, 101, 103, 114, 118, 119, 124, 133, 143, 158, 164, 166, 167, 176
private, xiv, 10, 25, 28, 29, 34, 59, 75, 89, 105, 177, 179, 189
private investment, 89
private sector, 10, 177
private-sector, 29, 34
producers, xii, 82, 87, 99, 100, 103, 107, 137, 141, 160, 164, 169, 180, 184, 185, 186, 187, 188, 189
production, xi, xii, xiv, 3, 27, 50, 51, 55, 57, 63, 64, 66, 73, 78, 81, 82, 83, 85, 86, 87, 89, 90, 96, 98, 99, 100, 103, 104, 107, 108, 145, 147, 150, 153, 155, 179, 180, 184, 189
production costs, 103, 107
production technology, 55, 89
profits, 119, 141
proliferation, 18, 35, 40
propulsion, 12, 38
protection, viii, 2, 10, 11, 23, 184
protocols, 137
proxy, 158
public, xi, 9, 10, 26, 29, 32, 43, 50, 67, 79, 86, 89, 95, 119, 135, 155, 187
public education, 119
public finance, 135
public interest, 95
public policy, 79, 155
public safety, 86
public support, 9, 187
pyrolysis, 83, 84, 89, 90

Q

quotas, 104, 105

R

R&D, v, vii, viii, 1, 2, 4, 5, 6, 7, 8, 9, 10, 11, 12, 13, 36, 102, 108, 115, 176
radiation, 23, 30, 31
Radiation, 45
radioactive isotopes, 18, 36, 39
radioactive waste, ix, 15, 18, 19, 24, 27, 32, 78
rainforest, 130
range, 70, 132, 151, 155, 158, 159, 160, 173, 174, 175
rangeland, xiv, 179, 189
raw material, 107
raw materials, 107
reading, 45, 78, 174
Reagan Administration, 6, 40
reality, 122
reasoning, 143, 171
rebates, 160, 164
recognized tribe, 189
reconciliation, 186, 190
recovery, 26
recreational, 155
recycling, 24, 35, 36, 37, 158, 163, 176
redistribution, 158
refineries, 57, 100, 121, 150, 151
refinery capacity, 63
refining, 94, 99
Reform Act, 189
refrigeration, 153
regional, xi, 33, 40, 41, 42, 50, 67, 118, 184, 185, 189
regular, 62
regulation, 10, 29, 127, 158
regulations, xii, 23, 43, 79, 94, 121, 153, 174
regulators, 31
Regulatory Commission, ix, 5, 15, 17, 21, 45, 46, 47, 67, 78
regulatory requirements, 185
relationship, 75, 127, 130
renewable energy, xi, 50, 54, 55, 73, 164
renewable resource, 73
reprocessing, x, 16, 18, 27, 28, 29, 35, 36, 37, 44
research and development, vii, viii, xii, 1, 2, 24, 81, 82, 88, 98, 99, 102, 108, 177
Research and Development, 4, 5, 12, 13
residential, 51, 52, 56, 69, 70, 72, 110, 146, 150
residues, xi, 81, 82, 83, 86
resources, 10, 37, 52, 54, 55, 88, 119, 180, 187
retail, 62, 69, 79, 99, 110
revenue, xii, 82, 87, 88, 119, 124, 126, 135, 139, 140, 144, 148, 158, 159, 161, 162, 163, 164, 173, 176, 177
RFS, 101
Richland, 46
rights-of-way, 34
risk, 10, 28, 34, 38, 44, 86, 99, 101, 137, 142, 167, 172, 181, 187
risks, 28, 32, 133, 156
Romania, 56
rural, 90, 107

S

S&T, 35
safety, 26, 38, 41, 124, 125, 151, 152, 153, 167
sales, 75, 98, 106, 110, 136, 189
salt, 39, 42, 43
salt domes, 39
Savannah River Site, 37
scientific understanding, 132, 188
seabed, 37, 38
search, x, 16, 17, 28, 30, 39, 40, 41, 55
secret, 3
Secretary of the Treasury, 177
security, xiii, 28, 33, 44, 94, 96, 97, 98, 119, 164
self-employed, 177
self-interest, 141
Senate, xii, 23, 32, 33, 36, 38, 40, 45, 46, 47, 93, 95, 97, 109, 110, 111, 115, 140, 182, 190
Senator Reid, 23

separation, x, 16, 35, 44
series, 2, 86, 88, 161
services, iv, 4, 20, 119, 120, 155
shares, vii, 2, 189
shipping, 17, 25
shock, vii, 1
short supply, 66
shortage, 63, 68
shortages, 57
short-term, xiv, 75, 118, 125, 130, 133, 138, 161, 167
Short-term, 130
sign, 20, 30, 184
signals, 95, 155, 170
singular, 9, 11
sites, ix, x, 3, 16, 17, 19, 20, 24, 26, 27, 28, 30, 31, 32, 33, 34, 36, 37, 39, 40, 43, 44, 153, 181
Slovakia, 56
Smithsonian Institution, 12
smokestacks, 148
SO2, 130, 133, 136, 141
social costs, 119
social security, 119, 164
Social Security, 177
soil, xi, xii, xiv, 81, 82, 83, 84, 85, 86, 87, 89, 90, 179, 180, 188
soil erosion, 85, 180, 188
soils, 85, 90, 150, 152, 161
solar system, 38
South America, 90
South Carolina, 37
sovereignty, 143
Soviet Union, 4, 56
Soyuz, 12
species, 181, 188
spectrum, 75, 145
speech, 13, 47, 173
speed, 41, 106, 110
spent nuclear fuel, ix, 15, 18, 19, 33, 43
Sputnik, 4, 9, 12
SRS, 46
stability, 94, 126
stages, xii, 21, 82, 85, 144
stakeholders, 101, 103, 137, 141, 164, 173
standards, xii, 38, 41, 79, 94, 97, 98, 101, 108, 114, 115, 120, 174, 188
Standards, 45, 75, 80, 114, 172, 176, 180
starch, xiii, 94, 96, 103
State of the Union, xiii, 94, 108
State of the Union address, 108
state regulators, 31
statistics, xi, 50, 74, 78, 79
statutory, xiii, 33, 41, 95, 104, 117, 189
statutory obligation, 104
steel, 150, 152, 153, 160, 174, 175, 177
steel mill, 150
stigma, 139, 167, 170
stock, 129, 130, 138, 168
storage, x, 16, 17, 18, 19, 20, 24, 25, 27, 28, 30, 31, 32, 33, 34, 38, 39, 43, 44, 68, 84, 101, 187
streams, 119, 184
strength, 134, 135
subsidies, 135, 144
substances, 101, 150, 153
substitutes, 153, 174
substitution, 142, 150, 169
sulfur, 129, 133, 141, 170
sulfur dioxide, 129, 133, 141
summer, 64, 94, 95, 98
supplemental, 27
supplements, 20
suppliers, 95, 98
supply, vii, x, 35, 49, 55, 61, 63, 68, 73, 77, 86, 94, 95, 99, 101, 102, 114, 176, 177
supply disruption, 61, 77, 99
Supreme Court, 26, 121
surface water, 85, 180, 184
surprise, 155
sustainability, 86
synthesis, 102

T

takeover, 31
targets, 120, 138, 190
task force, 24
taste, 101
tax base, 148, 149, 153, 170

tax credit, xiii, 94, 95, 96, 97, 99, 106, 107, 110, 114, 115, 135, 143, 147, 149, 151, 152, 153, 160
tax deduction, 96, 162
tax incentive, xii, 93, 98, 99, 100, 103, 106, 107, 114, 115
tax incentives, xii, 93, 98, 99, 100, 106, 107, 114, 115
tax increase, 167
tax policy, 140, 159
tax rates, 65, 119, 143, 144, 154, 156
tax reform, 140
Tax Relief and Health Care Act, 103, 104
tax system, 125, 143, 144, 149, 151, 156, 158, 165
taxation, 136, 144, 145, 148, 149, 152
taxes, 62, 64, 69, 119, 132, 136, 141, 144, 148, 149, 154, 158, 159, 160, 161, 162, 164, 172, 173, 176, 177
taxpayers, 107
technical assistance, xiv, 179, 181, 185
technological change, 165
technological developments, 98
Technology Assessment, 47
technology transfer, 88, 184
Tennessee, 3, 40
terrorist attack, 30
Texas, 68, 183, 184, 190
thorium, 36
threat, viii, 2, 6, 9, 11, 167, 172
threatened, 9, 32, 63, 187
threats, 9
Three Mile Island, xi, 33, 50, 67
time frame, 19, 37, 130
time periods, 85, 130, 133
title, 29, 34, 36
tolerance, 184
total costs, 167
total energy, x, 49, 50, 52, 54
toxic, 101, 115
toxic substances, 101
toxicity, x, 16
trade, xiii, xiv, 79, 113, 117, 118, 120, 121, 123, 125, 126, 130, 132, 133, 134, 135, 136, 137, 138, 139, 140, 141, 142, 143, 144, 145, 152, 158, 160, 162, 164, 165, 166, 167, 170, 171, 172, 173, 174, 177
trade-off, 126
trading, 63, 87, 88, 136, 142, 176
training, 181
transfer, 29, 88, 184
transition, 164
transparency, xiv, 118, 135, 136, 166
transparent, 87, 135, 144
transport, 34, 60
transportation, x, xi, 5, 20, 28, 32, 49, 50, 51, 52, 54, 57, 59, 60, 70, 72, 75, 94, 96, 98, 102, 109, 112, 115, 146, 147, 150, 152
Treasury, 20, 125, 177
trees, 96, 97, 102, 181
tribal, 34, 185
trucks, 59, 75, 98, 120

U

U.S. Department of Agriculture, xiv, 83, 88, 89, 91, 98, 114, 151, 152, 179, 180
U.S. Department of Agriculture (USDA), 83, 89, 98
U.S. economy, 133, 145
U.S. Geological Survey, 152
uncertainty, 36, 43, 123, 124, 126, 131, 133, 154, 155, 158, 166
UNFCCC, 88
unforeseen circumstances, 28
United Kingdom, 173
United Nations, 88
uranium, 3, 27, 35, 36
uranium enrichment, 3
USDA, xiv, 83, 89, 91, 98, 101, 114, 179, 181, 182, 183, 185, 186, 187, 189, 190
USSR, 9
Utah, x, 16, 25, 34, 35, 46

V

values, 131, 132, 133, 155, 167
variables, 122, 131

vehicles, xii, xiii, 75, 93, 94, 95, 96, 97, 98, 104, 105, 106, 107, 108, 109, 110, 112, 114, 115, 153
Venezuela, 55
Vice President, ix, 15, 17
Vietnam, 90
volatility, xiii, xiv, 95, 118, 123, 124, 133, 135, 138, 166, 167, 171, 172

W

waste disposal, 26, 29, 36, 43
waste management, ix, 16, 24, 27, 28, 29, 34, 38, 181, 187
Waste Package, 46
wastes, x, 16, 26, 38, 102
water, xiv, 10, 19, 30, 31, 35, 36, 85, 90, 101, 108, 155, 179, 180, 181, 184, 188
water quality, 184, 188
water resources, 180
water supplies, 155
watershed, 185, 188
Weathered, 90
web, 7, 8, 12
Western Europe, 56
Western Hemisphere, 55, 56
wetlands, 188
wholesale, 79, 107
wildlife, 188
Wildlife Habitat Incentives Program, 189
wind, xi, 50, 51, 73
Wisconsin, 26, 45
workers, 164, 177
World Resources Institute, 173
World War I, 3, 9, 12
World War II, 3, 9, 12

Y

yield, vii, viii, xiv, 1, 2, 82, 102, 118, 125, 127, 130, 143, 156, 161, 165, 166, 171
Yucca Mountain, v, vii, ix, x, 15, 16, 17, 18, 19, 20, 21, 22, 23, 24, 25, 26, 27, 28, 30, 31, 32, 33, 36, 37, 39, 40, 41, 43, 44, 45, 46, 47

Z

zirconium, 31